The Geometry
of the Universe

K&E Series on Knots and Everything — Vol. 71

The Geometry of the Universe

Colin Rourke

University of Warwick, UK

World Scientific

NEW JERSEY · LONDON · SINGAPORE · BEIJING · SHANGHAI · HONG KONG · TAIPEI · CHENNAI · TOKYO

Published by

World Scientific Publishing Co. Pte. Ltd.

5 Toh Tuck Link, Singapore 596224

USA office: 27 Warren Street, Suite 401-402, Hackensack, NJ 07601

UK office: 57 Shelton Street, Covent Garden, London WC2H 9HE

British Library Cataloguing-in-Publication Data
A catalogue record for this book is available from the British Library.

Series on Knots and Everything — Vol. 71
THE GEOMETRY OF THE UNIVERSE

ISBN 978-981-123-386-9 (hardcover)
ISBN 978-981-123-387-6 (ebook for institutions)
ISBN 978-981-123-388-3 (ebook for individuals)

For any available supplementary material, please visit
https://www.worldscientific.com/worldscibooks/10.1142/12195#t=suppl

Desk Editor: Yumeng Liu

Foreword

The universe in which we live is a wonderful, amazing place. We know a great deal about it through a series of fantastic observations using ground and space based telescopes and other equipment, including recently, equipment to detect gravitational waves. Cosmology, the study of the universe, arouses a great deal of public interest, with serious articles both in the scientific press and in major newspapers, with many of the theories and concepts (eg the "Big Bang" and "black holes") discussed, often in great depth. The observations that support these discussions use sophisticated and expensive machinery both on the earth (eg the LIGO equipment used to detect gravitational waves) and in orbit around the earth (eg the Hubble telescope). This is very big science with budgets to rival those of whole countries.

There is a consensus for the theoretical framework supporting and interpreting these observations, which is known as the "standard model". This starts with a singular event known as the "Big Bang" and expands from there to fill the entire visible universe. The image presented is of a complete and full theory that (with a few minor unsolved problems) explains all the observations that we have.

This image is (like many images) completely false. There are huge problems with the standard model and is the aim of this book to present an alternative model, which avoids the most glaring of these problems. The new model is based around a geometric description of the universe and hence the title. In this new model there is no Big Bang — indeed the model has what is known as the Perfect Copernican Principle that there should be no special times or places — and no dark matter, which is replaced by a simple geometric hypothesis. Cosmology is a highly technical subject involving some deep and difficult mathematics and presents a challenge to any author attempting to communicate in a way that is accessible to a non-technical audience. Nevertheless I shall attempt to do this. But I will also allow myself the luxury of giving full technical details for readers who wish to explore them.

Accordingly the book is divided into three parts:

Part 1 The whole story presented as far as possible for a nontechnical reader

Part 2 The same story, told again but for a reader with some technical knowledge

Part 3 Appendices with full technical details of several of the important topics covered.

Part 1 should be readable (and understandable) by anyone with a nodding acquaintance with the basic language of cosmology: events, lights paths, galaxies, black holes and so on. It covers the whole story of the book in as untechnical way as possible given the scope of the topics covered.

Part 2 covers the same ground again but with enough technical details to satisfy a reader with basic knowledge of mathematics and/or physics.

Part 3 consists of appendices which are referred to in the other parts and which also contain the highly technical material omitted from Section 2.

Several parts of this book are based on joint work with Robert MacKay and I thank him for allowing me to use this material and also for a thoughtful critical read. I would also like to thank Rosemberg Toala Enriques for the use of the material in the draft three author paper [88] on quasars. Special thanks are due to Robert MacKay, Ian Stewart and Rob Kirby for unfailing support through the discouraging process of attempting to publish this work in serious scientific journals.

An early version of this book with the title "A new paradigm for the universe" (suitable for readers with technical knowledge) is stored on the arXiv at `astro-ph/0311033` and can also be downloaded from the author's web page.

Colin Rourke
October 2020

Mathematics Institute, University of Warwick, Coventry CV4 7AL, UK
`cpr@msp.warwick.ac.uk`   `http://msp.warwick.ac.uk/~cpr`

AMS classification   85A40; 83C57, 85A15, 85A05, 83F05

About the author

Colin Rourke is a professor emeritus of mathematics at the University of Warwick, and has also taught at the Princeton Institute for Advanced Study, Queen Mary College London, the University of Wisconsin at Madison, and the Open University, where he masterminded rewriting the pure mathematics course; he has recently retired from lecturing after completing a half-century (of 50 years of lecturing). He has written papers in high-dimensional PL topology, low-dimensional topology, combinatorial group theory and differential topology.

In 1996, dissatisfied with the rapidly rising fees charged by the major publishers of mathematical research journals, Colin decided to start his own journal, and was ably assisted by Rob Kirby, John Jones and Brian Sanderson. That journal became Geometry & Topology. Under Colin's leadership, GT has become a leading journal in its field while remaining one of the least expensive per page. GT was joined in 1998 by a proceedings and monographs series, Geometry & Topology Monographs, and in 2000 by a sister journal, Algebraic & Geometric Topology. Colin wrote the software and fully managed these publications until around 2005 when he cofounded Mathematical Sciences Publishers (with Rob Kirby) to take over the running. MSP has now grown to become a formidable force in academic publishing. With his wife Daphne, he runs a smallholding in Northamptonshire, where they farm Hebridean sheep and Angus cattle.

In 2000 he started taking an interest in cosmology and published his first substantial foray on the ArXiv preprint server in 2003. For the past twelve years he has collaborated with Robert MacKay also of Warwick University with papers on redshift, gamma ray bursts and natural observer fields. He now feels that he has mastered the basis of a completely new geometrical way of understanding the universe, without either dark matter or a Big Bang. It is this that is presented in this book.

Photo: Pip Sheldon

Contents

PART 1

A nontechnical overview

Chapter 1

From the Greeks to Einstein

Many ancient civilizations studied the heavens and plotted the positions of the visible planets. The belief that heavenly bodies such as planets affect humans was widespread, and both ancient Greek and Roman civilizations used the same words for some of their Gods as for the planets. This book is not concerned with the possibility that there might be some substance in these beliefs (Astrology), but rather with the nature and dynamics of the heavenly bodies themselves (Astronomy) and the ultimate nature of the entire universe (Cosmology). It is however a salient fact that belief in astrology powered much early astronomy leading to the construction of observatories such as the Ziggurats of the Babylonians and Stonehenge.

Figure 1.1: Ziggurat at Ali Air Base Iraq (left) and Stonehenge (right)

But this is not a history book. There are excellent accounts of these ancient observations and speculations to be found in the literature and little more will be said about them in this book, which is concerned with modern astronomy. This starts with the publication by Copernicus [20] just before his death in 1543 of his book *On the Revolutions of the Celestial Spheres*. Or perhaps it really started with his ancient precursor Aristarchus of Samos (c 310 – c 230 BC) who, in about 250 BC, formulated the same theory as Copernicus, namely that the sun (the central fire according to an earlier partial precursor, Philolaus) is the centre of the solar system [19, 26]. Aristarchus had the correct order for the planets and also speculated (correctly) that stars are distant suns and therefore that the universe might be very much larger than it seems at first sight.

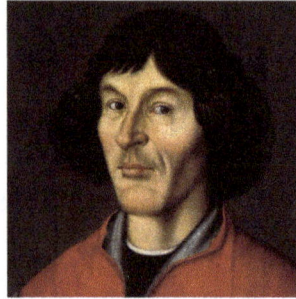

Figure 1.2: Aristarchus of Samos (left) and Nicolaus Copernicus (right)

However it is a sad fact that Aristarchus' insights were forgotten, or more probably ignored, in favour of the more "obvious" view that the earth is the centre of the universe. The dominant theory originating with Ptolomy and embodied in the teachings of Aristotle was that the earth is surrounded by a series of concentric crystal spheres which carry the planets and rotate to create the observed orbits. The outermost sphere is the sphere of the fixed stars, which was supposed to be eternal and unchanging.

There is a cautionary tale here: being correct does not imply being recognised as correct. The fixed Aristotelian view of the universe dominated thought and buried Aristarchus' insights until Copernicus brought them back into the light. The Roman Catholic church and in particular the Jesuit order imposed the Aristotelian view as divine doctrine and it was this that caused Gallileo so much trouble.

The same story (accurate insights and observations buried and forgotten) repeats again and again and there will be many more examples later in the book.

1.1 Kepler and Newton

Serious mathematical cosmology starts with Kepler and Newton (Figure 1.3).

Figure 1.3: Johannes Kepler (left) and Isaac Newton (right)

Kepler worked with the gifted astronomical observer, Tycho Brahe, and continued his work after his death. The observations of these two were sufficiently detailed and accurate for Kepler to formulate his three laws of planetary motion.

Kepler's laws of planetary motion

1 *The orbit of a planet is an ellipse with the Sun at one of the two foci.*

2 *A line segment joining a planet to the Sun sweeps out equal areas in equal intervals of time.*

3 *The square of a planet's orbital period is proportional to the cube of the length of the semi-major axis of its orbit.*

These laws informed and inspired Newton in the formulation of his general theory of gravitation in which the laws become theorems with rigorous proofs.

It is worth examining the second law in detail, because this law is equivalent to the law of conservation of angular momentum which will play an important role in several topics in this book. These include a geometrical explanation for "dark matter" and a theory of quasars which is compatible with observations of Halton Arp, which are ignored (or denied) in current mainstream cosmology. But first here is an overview of Newton's laws of motion and theory of gravitation.

1.2   Newtonian physics: gravitation and dynamics

Newton assumes an infinite universal frame of reference with three coordinates of space and one of time. The geometry used in this frame is ordinary (school) Euclidean geometry. There is no a priori justification for such a frame and Newton effectively says that it is "God-given" and refers to it a God's sensorium.

Ernst Mach severely criticised this assumption and his ideas influenced Einstein in his formulation of relativity: this will be discussed at length later (see Chapter 4).

Newton's laws of motion

1 (Conservation of momentum)   *An object either remains at rest or continues to move at a constant velocity in a straight line, unless acted upon by a force.*

2 (Change of momentum under the action of a force)   *The acceleration of a body acted on by a force is proportional to the force and in the direction of the force.*

The constant of proportionality is the mass m of the object and this law gives the familar equation $F = ma$, where F is the applied force and a the resulting acceleration.

3   (Action and reaction equal and opposite)   *When one body exerts a force on a second body, the second body simultaneously exerts a force equal in magnitude and opposite in direction on the first body.*

Implicit in law 1 is the concept of an *inertial frame*. This is a coordinate frame (three dimensions of space and one of time) which might just cover a small part of the universal frame, which is either at rest or moving with uniform velocity in a straight line with respect to the universal frame. Newton's laws hold in any inertial frame but not in any other frame, so that an inertial frame can be defined as a frame in which the laws hold. This may seem like a superfluous concept at this point, but it will become important when discussing relativity.

Newton's law of gravitation

Every particle attracts every other particle in the universe with a force that is directly proportional to the product of their masses and inversely proportional to the square of the distance between their centers.

As an equation the law takes the form $F = G\frac{m_1 m_2}{r^2}$, where F is the attractive gravitational force between two objects, m_1 and m_2 are their masses, r is the distance between the centres of the masses and G is Newton's gravitational constant.

1.3   Derivation of Kepler's laws 2 and 3

As remarked earlier, Kepler's laws of planetary motion become provable theorems under Newton's laws of motion and gravitation. The proof of the first law is rather more technical than the others and, as nothing depends on it, will be omitted; it can be readily found by searching on the web. Law 2 (equal area law) is essentially the law of conservation of angular momentum, a concept central to much of the book, and has a short elementary proof (due to Newton) using a little school (Euclidean) geometry. Law 3 is really just the inverse square law.

Note to readers This part of the book is supposed to be nontechnical, but occasionally contains some technical material, such as the proofs of Kepler's laws 2 and 3 below. These proofs are included for interest and can be omitted with little loss.

Angular momentum

Suppose a particle P of mass m is moving smoothly in a plane Π and O is a point in that plane. Think of Π as the plane of the solar system, P as a planet and O as the sun, if you like. The motion of P is along a straight line ℓ to first order. The *angular momentum* of P about O is defined to be the product mvd of the mass m of P with the velocity v of P along ℓ and with the distance d of O to ℓ (Figure 1.4). And this is $2m$ times the rate at which the line joining O to P sweeps out area at any point of ℓ. Two points (P and P') are illustrated in the figure and the two swept out triangles (OPX and $OP'X'$) have the same area $vd/2$ (equal bases v and the same height d).

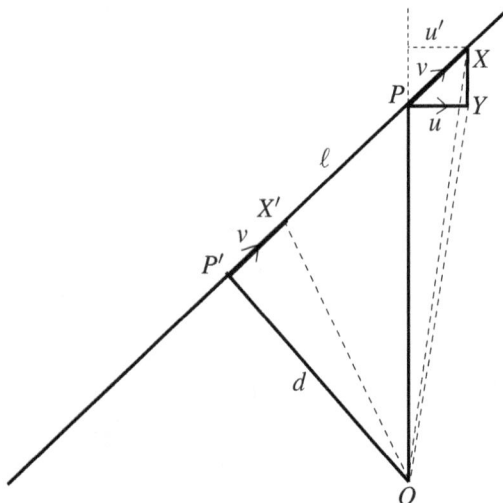

Figure 1.4: Proof of conservation of angular momentum

Angular Momentum of P about O is 2m times the rate at which the line joining P to O sweeps out area.

Thus law 2 is the same as the conservation of angular momentum.

Now resolve v into *tangential* motion u perpendicular to the line to O, and *radial* motion along the line to O. The figure also shows that the tangential motion of P (ie ignoring the radial motion) has the same angular momentum. (Triangles OPX and OPY have the same area because they have the same base OP and equal heights u and u'.)

Angular Momentum depends only on the tangential velocity of P.

Proof of law 2 This is where it gets clever. So far we have not used the fact that P is moving under the effect of the central attraction from O. Think of the attraction as a series of radial impulses. Each impulse affects only the radial velocity and leaves tangential velocity unchanged and hence, by what we have just proved, leaves angular momentum unchanged. Going to the limit when the impulses become a continuous force, proves the law.

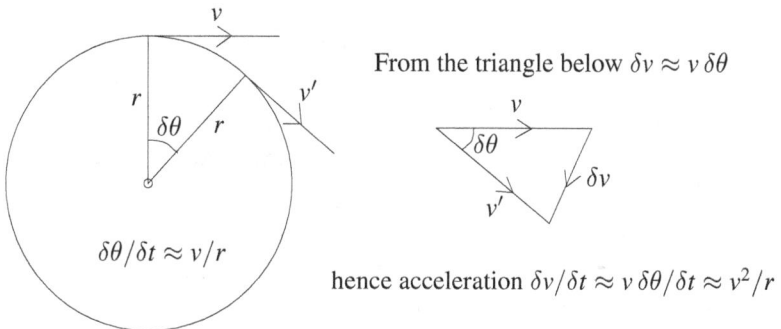

From the triangle below $\delta v \approx v \, \delta\theta$

$\delta\theta/\delta t \approx v/r$

hence acceleration $\delta v/\delta t \approx v \, \delta\theta/\delta t \approx v^2/r$

Figure 1.5: Inward acceleration for a circular orbit

Proof of law 3 For simplicity assume orbits are circles. Figure 1.5 shows that the inward acceleration is v^2/r and this is equal to MG/r^2, where M is the mass of the sun, by Newton's second law of motion and the law of gravitation. In other words $v^2 = MG/r$. But period is $2\pi r/v$ so period squared is $4\pi^2 r^2/v^2 = r^3(4\pi^2/MG)$, ie the period squared is proportional to the radius of the orbit cubed, which is law 3.

1.4   Maxwell and the road to Special Relativity

Figure 1.6: James Clerk Maxwell (left) and Henri Poincaré (right)

Newton's comprehensive remodelling of Physics around the three laws of motion and the law of gravitation ushered in a golden age of physics where all problems seemed accessible to the new methods. The universe seemed to be entirely explained at least in principle and this fitted well with the age of enlightenment. In parallel developments, great strides were made in understanding the phenomena of electricity and magnetism culminating in the epochal work of James Clerk Maxwell (Figure 1.6 left) who unified optics, electricity and magnetism in the single theory of electromagnetism. His equations give a complete formulation of the theory. Featured in the equations is a new physical constant c, the velocity of electromagnetic radiation in a vacuum. Since one of the key pillars of the theory is that light is electromagnetic radiation, this constant is usually called "the speed of light".

The æther and the Michelson–Morley experiments

Maxwell assumed that light (and other electromagnetic radiation) is a wave, like a wave in the ocean, and the medium in which the wave travels is the "luminiferous æther". Since the earth is also travelling in the æther and varies its velocity at different points in its orbit around the sun, it should be possible to measure a variation in the speed of light by suitably sensitive instruments. The famous Michelson–Morley experiment sought to do this but failed. Several later experiments were conducted which ought to have detected the motion of the earth through the æther but also failed [25]. These experiments precipitated a crisis in physics since they seemed to show that any observer, whether in motion or not, measures the same speed for light. This is impossible under Newtonian physics with its universal measure for both space and time. Suppose two observers A and B, with A travelling towards B, measure the speed of the same flash of light L which is also travelling towards B. Suppose A is approaching B with velocity v and measures L having velocity c. Then A will measure the velocity of L towards B as $v + c$ and if A and B agree on measurements for space and time then B will also measure the light to have velocity $v + c$ contradicting the Michelson–Morley observations (Figure 1.7).

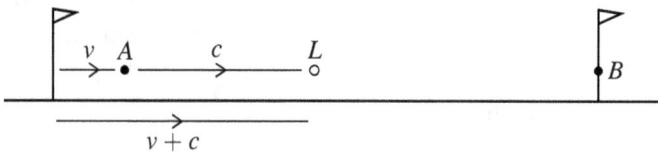

Figure 1.7: *Newtonian picture*   The posts are fixed in B's space. A is moving at velocity v towards B and L is moving at c towards B relative to A. Both observers see L as moving at $v + c$ towards B.

The inescapable conclusion is that space and/or time are *not universal* but depend on the velocity of the observer. The motion of an observer causes lengths to contract or time to dilate or both. Several mathematicians worked to resolve this problem and the solution, *special relativity*, was found in essence by Henri Poincaré (Figure 1.6 right), Hendrik Lorentz and Hermann Minkowski (Figure 1.8), as well as (famously) Albert Einstein (Figure 2.1). There is nothing suspicious about the same theory being discovered by several researchers at the same time. In this case the problem was well known: the apparently paradoxical results of the Michelson–Morley experiments and the need to drastically modify Newtonian physics to make it consistent with these experiments. Moreover the mathematics

Figure 1.8: Hendrik Lorentz (left) and Hermann Minkowski (right)

involved is not deep or particularly subtle. Appendix A contains a fairly easy derivation of special relativity based on the observed fact that all observers see light paths the same way, and the principle that all inertial observers (moving with constant velocity in a straight line) should be equivalent (Einstein's *relativity principle*). Therefore it is not surprising that several mathematicians / physicists arrived at essentially the same theory at more-or-less the same time.

At this point the reader is recommended to turn to Appendix A for an introduction to the basic concepts of relativity. Read as far as is comfortable. Try to cover at least the sections on special relatvity (up to Section A.5). A quick summary of the basic concepts needed to proceed comes at the start of the next chapter.

Chapter 2

Einstein, relativity, model building and de Sitter

Figure 2.1: Albert Einstein lecturing in 1921 and with his first wife Mileva

Welcome back from reading the appendix on relativity. A quick summary of the important concepts and facts that you should have picked up are numbered 1–6 below:

1 Events An event is a point of *space-time*, in other words it has definite position (space) *where* it happens and a definite time *when* it happens.

2　Causality　Events A and B are causally related if it possible for a signal (travelling at speed less than or equal to c) to pass from A to B. In other words A *precedes* B.

3　Observers and world-lines　An observer or world-line is a path in space-time with each point preceding later points.

4　Relativity of space and time　Space and time are relative concepts which depend on the observer.

5　Constancy of speed of light　The speed of light (measured in a vacuum) is the same for all observers.

6　Inertial frames and observers　Frames in which Newton's laws hold to first order and observers stationary in such frames. Inertial frames exist and to first order the geometry of an inertial frame is the same as Minkowski space.

The final two items were not covered in the appendix and are included for completeness, the first characterises special relativity and the second, general relativity. More will be said about them in this chapter.

7　Einstein's special principle of relativity　The laws of physics are unchanged by uniform relative motion.

8　Einstein's general equivalence principle　All systems of reference are equivalent with respect to the formulation of the fundamental laws of physics.

2.1　Special Relativity

Einstein based his derivation of special relativity [40] on two of the items above:

7　The special principle of relativity

5　The constancy of speed of light

He made no assumption about the existence or properties of an æther. [The derivation of the basic properties of special relativity given in Appendix A uses essentially the same assumptions and covers similar ground.]

As mentioned in the last chapter, other mathematicians and physicists (in particular Poincaré, Lorentz and Minkowski) also found the basic properties of special relativity. Einstein's treatment is elegant but not especially superior to the other treatments. The pre-eminence of Einstein for the theory of relativity would not have been recognised if he had not gone very much further and developed a far more general theory.

2.2 General Relativity

Einstein's aim is well summarised in the *equivalence principle* given as item 8 above. He wanted *all* observers to be equivalent, not just inertial observers. This means that an observer affected by a gravitational force should be equivalent to one being accelerated (and hence subject to a virtual force equivalent to the acceleration). One example is often used: an observer in a lift experiences a force that accelerates the lift upwards (or downwards) and this should be identical for all experimental purposes to the effect of increasing or decreasing the gravitational pull of the earth. He referred to this property as "general covariance".

The idea of pure genius that he had, was that the force of gravity is due to curvature of space-time and similarly the forces due to (or causing) acceleration must also be encoded in the local space-time geometry. From this idea to the final form of the theory was not an easy step, and Einstein spent the time from 1905 to 1915 on this task. He wasted the last two years on a false trail [13], and only finished the theory because of fear that David Hilbert, Figure 2.2 (left), might publish a similar theory before he did. Incidentally there is a false idea circulating that Einstein was helped to the final form of his equations by Hilbert, see [28]. Like the theories that Einstein borrowed ideas from the other discoverers of special relativity [27] or that he effectively collaborated with his first wife Mileva [9], there seems to be no real substance to this. Einstein and Hilbert enjoyed a very fruitful correspondence and followed each other's research closely. There is no doubt that this correspondence spurred Einstein into finishing his theory, but this is not "help" in the usual meaning of the word!

There is a fairly simple exposition of general relativity in Section A.6 in the appendix and now might be a good moment for the reader to turn back to this. Here is short piece of motivation. Curvature of space-time is measured by a complicated 20 component tensor, the Riemann tensor. There is a much simpler measure of curvature, with only 12 components, namely the Ricci tensor which determines the way volume grows. If this is positive then nearby parallel geodesics will tend to converge (as if under the influence of a force). Thus curvature can "cause" space to behave as if there are forces acting. The formulation that Einstein eventually found after much effort was in terms of this, the Ricci curvature, rather than the more general Riemann curvature. There is information in the Riemann tensor unused by general relativity and Einstein (and others) spent a good deal of time and energy trying to use this spare information to extend the theory to other physical measurements such as electromagnetic fields. Einstein was still working on this when he was near to death but this program is still incomplete.

The idea of Einstein's formulation of general relativity is that it should say the curvature of space-time (in other words the "force" experienced by objects in space-time is equal to mass-energy-momentum (in other words the "matter" in space-time). The analogy is the Newton equation $F = ma$. The curvature tensor used is called the *Einstein tensor* and denoted G and it is close to the Ricci tensor. There is a *stress-energy* tensor T, which encodes the energy and momentum of matter, and the equations read $G = 8\pi T$. (The constant 8π is found by considering simple special cases.)

2.3   Model building

Having arrived at a satisfactory formulation for general relativity, Einstein started at once on model building — trying to find a good mathematical model for the universe as a whole within the framework of general relativity. He wanted various properties from this model and two important ones in particular:

(a)   it should be static,

(b)   it should embody Mach's principle.

Figure 2.2: David Hilbert (left) and Willem de Sitter (right)

A static solution seemed the natural thing to aim for. There was no reason to expect that the universe as a whole was anything but permanent and unchanging. The observations of Vesto Slipher, and Hubble–Humason [94, 59] which suggest that this expectation is false — the universe is apparently expanding — were yet to come.

Mach's principle will be explained shortly (Section 2.4).

Whilst model building, Einstein started working with the Leyden astronomer, Willem de Sitter, Figure 2.2 (right), and between 1916 and 1918 they conducted a strenuous correspondence (often called a "debate") about models for the universe. There is a very good account of this in Janssen [60]. The important outcome for the geometry of the universe presented in this book is that a perfect model for the universe (de Sitter space) emerged from this. The density of ideas bounced between these two physicists is much closer to collaboration than debate and the outcome should probably be seen a joint work rather than allocated to either in particular. Also two first class mathematicians, Hermann Weyl and Felix Klein, Figure 2.3, were closely involved with the final understanding of this new model, so perhaps it should more accurately be labelled *Einstein–De Sitter–Weyl–Klein space*!

Willem de Sitter first persuaded Einstein to abandon attempts to postulate boundary conditions at infinity as a means of expunging the remnant of

Figure 2.3: Felix Klein (left) and Hermann Weyl (right)

Newton's absolute space which remains in general relativity. Einstein responded by abolishing infinity! He proposed a spatially closed model with the simplest closed 3-manifold, namely the 3-sphere S^3, as the space. As a space-time, a static universe with the 3-sphere for space coordinates is $S^3 \times \mathbb{R}$, a cylinder with the 3-sphere for cross-section (rather than the circle for the usual cyclinder), Figure 2.4.

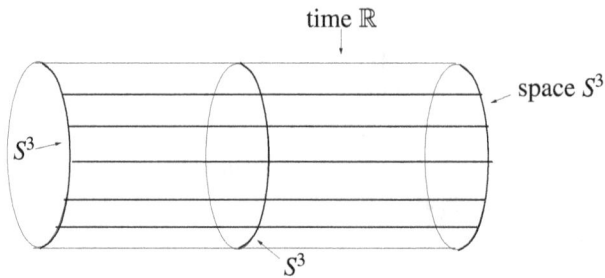

Figure 2.4: Einstein's cylindrical universe

Notice that this model disentangles space and time. It reinstates the universal time coordinate which fails so badly in special relativity. There will be more to say about this later.

Einstein's "biggest blunder"

Einstein immediately ran into a problem. To obtain a static solution to the equations for a cylindrical universe, he needed to add a *cosmological constant* κ times $\mathbf{g}$ to his tensor, so the equations become $G + \kappa \mathbf{g} = 8\pi T$ where $\mathbf{g}$ is the metric. After the observations of Slipher et al, which suggest that the universe is expanding, Einstein rescinded his cosmological constant κ calling it his "biggest blunder". If he hadn't introduced it, he could have predicted the observed expansion! Since the 1998 WMAP observations, most cosmologists are happy to keep the cosmological constant since the universe seems now to approximate de Sitter space which also has a positive cosmological constant.

From the author's point of view, the biggest blunder, not just of Einstein but of the whole community, was the reintroduction of a universal time in mainstream models for the universe in the large. There is no universal time in either special or general relativity. It is the assumption of a universal time that leads to the (false) Big Bang theory which dominates current cosmology. As remarked above, Einstein's cylindrical universe does have a universal time.

Incidentally, in 1930, Eddington showed that Einstein's cylindrical universe as originally planned is not stable [39] and therefore it does not provide the permanent static model that Einstein wanted. It is also an unhappy fact that Einstein's conviction that general relativity would automatically satisfy Mach's principle foundered on de Sitter space, as will be seen shortly.

2.4 de Sitter space

Whilst examining the calculations for Einstein's cylindrical model, de Sitter found another static model based on the 3-sphere, but with a horizon at spatial infinity. This model is not a cylinder (the space and time coordinates are intertwined) but it is a vacuum solution with, like Einstein's cylinder, a positive cosmological constant.

Two views of de Sitter space are drawn in Figure 2.5. Both show the full space as understood by Klein. It is a sphere in Minkowski 5-space (the

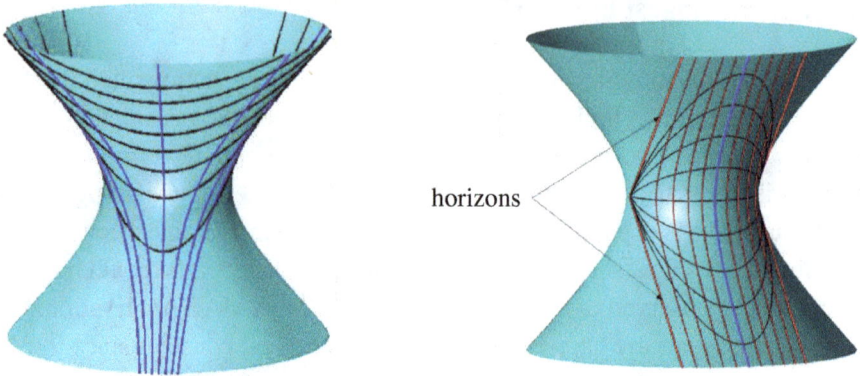

Figure 2.5: Figures reproduced from [77] of Klein's model of de Sitter space showing the central expanding frame (left) and the de Sitter static frame (right)

illustration is a sphere in Minkowski 3-space). This is a perfect symmetric model for the universe and is the natural first model to look at. Note that the metric is distorted in the pictures, in the same way that the metric is distorted in the pictures of Minkowski 2-space in Appendix A. The natural flow is along hyperbolas. Neither de Sitter nor Einstein understood this full picture until Felix Klein explained it in a letter written in June 1918. The space proposed by de Sitter is the region in the right hand picture bounded by the orange lines labelled "horizons". "Space" in de Sitter's metric is the horizontal circle (3-sphere in the full 5-dimensional picture) which flows upwards over time into the sloping ellipses. The orange lines appear to be singular — as horizons at spatial infinity. BUT the flow is not a natural time-like flow, under which points would flow along geodesics. Only the central (blue) flow line is a geodesic, so this flow could not be the flow of time in a real universe because of the torsion effects of the non-geodesic flow. Hermann Weyl also worked on understanding de Sitter's proposed model and patched pieces together to make a model similar to Klein's. But he left a zone of matter around the equator which allowed Einstein to discard this as a proof that his equations did not imply Mach's principle. Finally Klein convinced all the other three that de Sitter space is completely symmetrical and the "horizons" and equatorial singularities are just artefacts of a bad choice of coordinates. De Sitter space is a vacuum solution for Einstein's equations. Mach's principle relates matter and inertia. De Sitter space has inertia but no matter and Einstein finally accepted that Mach's principle was not automatically satisfied by solutions to his equations.

Before giving the promised explanation of Mach's principle it is worth remarking that, in the 5th edition of his book [103], Weyl showed that de Sitter space, as now understood by all the correspondents, exhibits expansion: any two geodesics which are not cofinal (ie which do not tend towards the same point at $+\infty$) eventually move apart at an exponential rate with velocity proportional to distance (usually called the "Hubble Law"). Figure 2.5 (left) shows the expanding frame in de Sitter space based on the central geodesic. The vertical blue hyperbolae are the sheaf of geodesics originating at the same point at $-\infty$ and the transverse black curves are flat 3-spaces in the 4-dimensional version (straight lines in the dimension illustrated in the picture) and there is a natural expansive flow along the blue curves. He tied this in with observations of Slipher [94], which preceded Hubble's observations. Weyl's book was published in 1923, four years before Lemaître's paper [64] covering similar ground and six years before Hubble's first paper on his law [58]. Note that Lemaître has now been recommended by the International Astronomical Union for credit in the formulation of the Hubble law which should now be called the Hubble–Lemaître law [11], perhaps it really ought to be called the Slipher–Weyl law! In view of this history and the IAU recommendation, this book proposes the following compromise naming convention.

Convention The "Hubble law" will now be called the "HLSW law" (Hubble–Lemaître–Slipher–Weyl law) and similar changes will be made to related phrases. For example Hubble's constant becomes the HLSW constant. This is in analogy to labelling the metric for the standard model, the FLRW-metric.

The subject of redshift in de Sitter space, and the space itself, will be covered in detail in Parts 2 and 3 (Sections 9.3–9.6 and Appendix B) where more details on all the points mentioned here can be found. It particular it will be seen that de Sitter space is *not globally expanding*. Expansion is a grand illusion and there is a dual contraction which balances the books.

de Sitter space and Mach's principle

Figure 2.6: Ernst Mach (left) and Dennis Sciama (right)

But to finish this chapter, here is a short account of Mach's principle. This principle will be discussed in depth in Part 2 (Chapter 4) and the formulation due to Dennis Sciama Figure 2.6 (right) is central to the solutions to the dark matter problem and quasar problems presented in this book.

Ernst Mach, Figure 2.6 (left), wrote a very influential book [69] severely criticising Newton's hypothesis of a universal frame as having no reality. He proposed that all assumptions about space and time should have their origin in observable quantities. There is nothing observable in Newton's infinite universal frame of reference. This principle is applied in particular to the inertial properties of space and time, which should be determined by the (theoretically observable) distribution of matter in the universe; for example, Berkeley [32] suggested that Newton's local rest frame could be defined by distant "fixed" stars.

The ideas in Mach's book influenced Einstein who thought that his theory, connecting inertial effects to matter via the curvature of space-time should automatically embody Mach's principle.

There are many effects that are observable which cannot be directly measured. The canonical example here is *rotation*. An observer can tell that he is rotating without leaving his closed windowless spaceship, because there are forces that he experiences (for example Coriolis force) that he does not experience if he is not rotating. But what possible difference is there between him rotating, and him being still with the universe rotating around him? The conclusion is that the forces he experiences are due to some mysterious effect of the rotation of the universe around him and this is also called "Mach's principle". These considerations lead to the philosophically compelling idea that the concept of acceleration or rotation must be connected to the main distant mass of matter in the universe. In other words, inertia is determined by the global distribution of matter in the universe. The local concept of inertial frame (recall: a frame in which there is no acceleration or rotation) is correlated with the distribution and motion of all the matter in the universe. Any theory which aims to accurately describe the universe must have such a property in some form.

But de Sitter space is a solution to Einstein's equations with no matter in it, so the local inertial frames (which exist as in all Lorentz manifolds) cannot be determined by the distant masses in the universe. So Einstein's hope that his equations would entail Mach's principle failed.

As remarked above, Mach's principle is central to important parts of this book: the solution to the dark matter problem, the dynamics of galaxies, and the way to understand Halton Arp's controversial observations of quasar growth. Since Mach's principle is not built in to General Relativity, it will be necessary to assume the principle as an extra hypothesis. Exactly what needs to be assumed is explained in detail in Chapter 4.

Chapter 3

The biggest blunder, dark matter and quasars

3.1 The biggest blunder

The biggest blunder in the development of cosmology was not Einstein's alone, and it was not concerned with the cosmological constant. It was made by the whole community in the course of the Einstein–de Sitter–Weyl–Klein debate and the aftermath and the model building. It was the discarding of de Sitter space as a possible model for the universe and the use instead of the "standard model" (3.2) with its universal time coordinate.

De Sitter space is a perfect symmetric model for the universe and is the natural first model to look at. It is a sphere. It may not look like a sphere, but that is because the metric is not seen accurately in ordinary space. It is a sphere in Minkowski space — a hyperbolic sphere. It has a huge amount of symmetry, but it was discarded as a suitable model for no cogent reason that can be found in the literature. It's almost as if it was always regarded as a curiosity and not a serious model.

Weyl's postulate

Weyl bears a huge burden of blame for this blunder. In [102] he proposed
a postulate that seems to have been widely accepted. It proposes that the
whole universe is the "domain of influence" of one particle moving on a
geodesic. This is illustrated in Figure 2.5 (left). The expanding frame filled
with blue geodesics is precisely the same as all the events (points) that can
be reached by a light signal starting on the central "home" geodesic and
this defines the domain of influence of this geodesic. Supposing that this
is the whole of the universe cuts the space down to exactly half. This is
best seen by noticing that in Figure 2.5 (right) the limit of the domain is
the pair of bottom horizons extended upwards. There are dual contracting
frames (for example the one based on the home geodesic is obtained by
turning Figure 2.5 (left) top to bottom). The contracting frame based on
the geodesic diametrically opposite to the home geodesic is precisely what
is removed when applying the postulate. Thus Weyl's postulate builds in
expansion as explained above, but discards the dual contraction which is
also built in to de Sitter space. There will be more to say about this duality
later in the book.

There are other formulations of Weyl's postulate and the reader is referred to
the Stanford Encyclopedia [15] for an overview.[1] The Encyclopedia quotes
Weyl as follows in justification for the postulate:

"The hypothesis is suggestive, that all the celestial bodies which we know
belong to such a single system; this would explain the small velocities of
the stars as a consequence of their common origin."

This is very weak as a justification. There were not many observations at
the time and the nearby stars that we see (and of which we can measure the
velocity) are in the same galaxy (the Milky way) as us and this accounts

---

[1]Equivalent formulations are that the world lines of all particles (presumably
galaxies) in the universe originate at the same point at $-\infty$, or equivalently that
the whole universe is the domain of influence of one particle, or equivalently that
the world lines of all particles form a sheaf orthogonal to a family of space-like
hypersurfaces, or equivalently that the smeared out matter in the universe is a
perfect fluid.

for their small velocities. We now know that there are galaxies with very large recessional velocities (high redshift) and it is suggested in [72] (cf Appendix F) that incoming galaxies with very high blue shifts are observed by us as gamma ray bursts. So there is a huge spread of velocities and no evidence from observations to support this justification.

One interesting fact about the postulate is not mentioned in the literature (to the author's knowledge). If the postulate is applied to a space, like de Sitter space, where the sheaf of world lines originating at a point at $-\infty$ squeezes down in backwards time by an arbitrarily large factor, then following particles down towards $-\infty$, after a finite time they must reach a point where they are so close to each other that they coalesce into one massive particle. In other words we have turned the clock back to a Big Bang singularity. Of course it is possible to imagine universes where this process converges to a non-zero cross-section. In other words it converges in backwards time to a universe rather like Einstein's cylinder. But perhaps we can assign to Weyl some of the responsibility for the worst hypothesis that has ever been made about the universe, namely the Big Bang theory. More on this anon.

De Sitter space as a model for the universe

It is now common ground that the universe is moving into a de Sitter phase. The main thesis of this book is that it has always been in a de Sitter phase. In fact de Sitter space perfectly fits all observations, not just the current phase. This thesis will be developed over the rest of the book. There is one problem with de Sitter space: it is a vacuum solution. A uniform density of matter could be added at the expense of altering the cosmological constant, BUT the matter in the universe is anything but uniform. Another principal hypothesis of this book is that standard (spiral) galaxies have supermassive black holes in their centres that drive the dynamic responsible for creating and maintaining the visible spiral structure. Mathematically it is a non-trivial problem to insert a black hole into a particular Lorentz manifold. This problem was discussed in MacKay–Rourke [71]. It seems difficult. But this is a purely technical problem. No one doubts that Lorentz manifolds can be made with mutiple black holes. It's just that the details of the construction have not been found. So this is no reason to discard de Sitter space as a possible model. There are other technical details that need attention concerning the contracting frames that exist in the uncut de Sitter space which we do not readily observe. There is an effect due to "observer selection bias" that makes us see the expansion far more easily than the dual contraction which is mostly seen in gamma ray bursts. There is also a horizon limiting visibility at about HLSW distance[2] (13.7 billion light-years) caused by the apparent background of gravitational disturbance in the universe which truncates geodesics and takes some of the fire out of these gamma ray bursts. This distance is the distance to the Big Bang in the current consensus model and is equally important in the model proposed in this book because the horizon at this distance is responsible for the Cosmic Microwave Background (CMB). It will usually called the visibility radius or visibility limit and the horizon, the visibility horizon.

The reader is directed to Sections 9.3–9.6 for more details here.

Being generous to Weyl, perhaps he calculated these large incoming velocities and made his postulate to rule them out. But it is far more likely that the postulate was made in an attempt to tame the wildness of full General Relativity, and return model making to a safer context. Having a universal time makes it possible to talk about the history of the universe as a whole. Without it all that can be described is the history of one particular particle (galaxy). But the price paid for this apparent simplification is exorbitant. The Big Bang is a hugely illogical hypothesis. It breaks all conservation laws. It hypothesises states of matter completely unknown. We can imagine Mach saying "I'd sooner believe in fairies at the bottom of my garden!"[3] The Big Bang is pure fantasy physics and it has needed shoring up by "inflation" which is another piece of fantasy physics and an obvious fudge factor!

---

[2]See the convention on page 21.
[3]An accurate quote from Mach is given by his son Ludwig from papers that he left: "... in my old age, I can accept the theory of relativity just as little as I can accept the existence of atoms and other such dogma" [69, page xiv].

3.2 The standard model

It is time to turn to the main model of current mainstream cosmology namely the *standard model*. Or perhaps we should say "standard models" because there are several models depending on choices in the construction. As mentioned several times, these models all have a universal time frame and therefore are at odds with one of the main points of relativity:

Space and time are relative concepts which depend on the observer.

They do all have the Copernican Principle: each space slice is assumed to be a 3-manifold of constant curvature and it is assumed that there are enough symmetries to carry any point on a given space slice to any other point on the same slice, so that the Copernican principle that no point is special is true. There are three main types of space slices: positive curvature, zero curvature (flat) and negative curvature. Examples of positive curvature include the 3-sphere and dodecahedral space, of zero curvature, ordinary 3-space and the 3-torus, and of negative curvature, the space for non-Euclidean geometry (hyperbolic space). But, because because all the models allow for time-dependent change, only the two static metrics, see the next paragraph, have the Perfect Copernican Principle that no event (point in space-time) is special.

The metric for the standard models (now called the FLRW metric) was first found by Friedmann [44] (1922) and then independently by Lemaître [64, 65] (1927) and again independently by Robertson [85] (1935–6) and Walker [101] (1937). Friedmann starts from Einstein's equations and proves that the only static metrics that fit the assumptions are Einstein's cylindrical model and de Sitter's static metric Figure 2.5 (right). Lemaître also starts from Einstein's equations and covers similar ground. Robertson and Walker prove a much stronger result: the FLRW metric is the only one on a spacetime that is spatially homogeneous and isotropic; this is a geometric result and is not tied to the general relativity. All four authors give metrics that start from a Big Bang singularity.

The current consensus model for the universe uses the FLRW metric and also starts from a Big Bang singularity and (as remarked earlier) includes an inflationary phase which was added to fix a serious problem with smoothing out the early universe. It is itself a serious problem!

It is time to list the other problems with the consensus model which will be tackled in this book.

Two philosophical problems

There are two major philosophical problems with the standard model. The first has already been mentioned: It is based directly on Einstein's theory of "General Relativity" (hereinafter referred to as EGR) which does not satisfy "Mach's principle" in general. Any theory which aims to accurately describe the universe must have such a property in some form. EGR does not.

The second problem is the (again philosophically compelling) principle known as the Copernican (or cosmological) principle. This is the principle that no particular location should be special. There is no centre to the universe: no "fingers of God". This should also be true of time as well as space, there should be no special times: we should not live in a special time any more than we live in a special place. This space-time non-speciality principle will be called the "Perfect Copernican Principle" or PCP (which can also be read as the "Perfect Cosmological Principle"). De Sitter space does have the PCP. There are symmetries that carry any point to any other point and further it can be arranged that any chosen time-like geodesic through the first point is carried to any chosen time-like geodesic through the second point. This is proved in Proposition B.1 in Appendix B. Obviously the consensus model does not satisfy this principle because it starts with a very special event, the Big Bang, taking place at a very special time.

Another early model with the PCP was the Bondi–Gold–Hoyle steady state theory (SST): this hypothesises that matter is created by empty space at exactly the correct rate to compensate for the observed HLSW expansion[4] — a hypothesis also briefly toyed with by Einstein (see [81]) as a possible alternative to (or explanation of) the cosmological constant, as a means of

---

[4]See the convention on page 21.

attaining a static solution to the field equations. Hoyle was an energetic proponent of this theory against the Big Bang theory (a sarcastic name that he invented) and his writings on the subject are well worth revisiting. Eventually he gave way because of the evidence of historic change in the composition of the universe from quasar observations — evidence that was in fact badly misinterpreted, see Section 3.6 and Chapter 6. Hoyle's SST could still be correct and in deference to his enthusiasm, a model with the PCP will be called a "Hoyle Universe". The model outlined in this book, based on de Sitter space, is indeed a Hoyle Universe, but it does not have the unnatural continuous creation hypothesis of the SST, which, like the Big Bang breaks commonly accepted conservation laws.

3.3 Dark matter

There are two other major problems with the current consensus model for the universe. The first (the so-called "dark matter" problem) is recognised as a major problem, whilst the second (the "Arp problem") is not recognised but ought to be. The dark matter problem arises from the observed rotation curves for galaxies. Typically the curve (of tangential velocity v against distance from the centre r) comprises two approximately straight lines with a short transition region. The first line passes through the origin, in other words rotation near the centre has constant angular velocity (plate-like rotation); the second is horizontal, in other words the tangential velocity is asymptotically constant, see Figure 3.1 (left). Furthermore, observations show that the horizontal straight line section of the rotation curve extends far outside the limits of the main visible parts of galaxies and the actual velocity is constant within less than an order of magnitude over all galaxies observed (typically between 100 and 300km/s) see Figure 3.1 (right).

Galactic rotation curves are so characteristic (and simple to describe) that there must be some strong structural reason for them. They are very far indeed from the curve obtained with a standard Keplerian model of rotation under any reasonable mass distribution. In a Keplerian model if v is asymptotically constant then the mass inside radius r is asymptotically equal to a constant times r and tends to infinity with r.

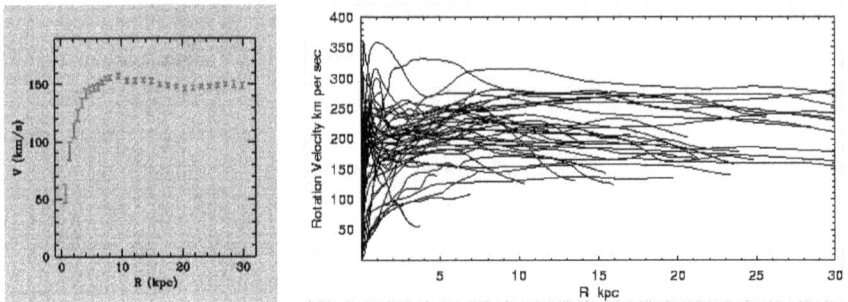

Figure 3.1: The rotation curve for the galaxy NGC3198 taken from Begeman [33] (left) and a collection of rotation curves from Sofue and Rubin [96] (right)

Nevertheless, in spite of the huge mass needed, a Keplerian model is exactly what is assumed in the consensus model. To square the circle, current theory hypothesises the existence of a huge amount of matter. Since this matter is not observed, it is called called "dark". It needs to be distributed in precisely the right way to make Keplerian rotation fit the rotation curve. This is extremely implausible for several reasons.

Firstly it has just been seen that the quantity of dark matter required is huge and tends to infinity with the radius of fit, which as mentioned above appears to be unbounded. Secondly it is unreasonable to suppose that exactly the right distribution of dark matter happened (by condensation) for every galaxy and thirdly, the final arrangement with most of the matter on the outside is dynamically unstable. For stability in a rotating system (such as the solar system or Saturn's discs) there must be a strong central mass to hold it together. Failing this the system will tend to condense into smaller systems. Finally despite the best efforts expended in the search, nor hair nor hide of dark matter has been found to date.

There is also a companion problem for the dynamics of spiral galaxies. The consensus model has no satisfactory model for galactic dynamics which explains the persistent spiral structure widely observed. The new model presented in this book solves the dark matter problem and gives a full model for galactic dynamics. It does this by building a limited version of Mach's principle into EGR (and this also deals with the first philosophical problem).

In essence all that is needed is to use a suitable relativistic model instead of a Keplerian one.

Sciama's principle and inertial drag

The solution to the dark matter problem given in this book uses a special case of Mach's principle due to Sciama. It concerns a relativistic effect: the dragging of inertial frames, henceforth abbreviated to "inertial drag" or ID. This is an effect whereby the non-inertial motion of a body (acceleration and/or rotation) causes the inertial frames at other points to be dragged. In his thesis [89] Sciama proposed that this effect should embody Mach's principle (that inertial frames depend on the total distribution of matter in the universe) and that this should specify the full dynamics of the universe. He illustrated this with a special case (a certain linear approximation to general relativity) where it all works perfectly. His formula for the ID effect of a rotating body is this:

Weak Sciama Principle (WSP) *A mass M at distance r from a point P, rotating with angular velocity ω, contributes a rotation of $kM\omega/r$ to the inertial frame at P where k is constant.*

It is called the "Sciama principle" to distinguish it from the more general "Mach principle" and "weak" because it specifies only the effect of one rotating body, instead of all non-inertial motions. The "Full Sciama Principle" (for rotation) states that the rotation of a local inertial frame is obtained by adding the effects for all rotating bodies in the accessible universe (ie not regressing faster than c).

The rotation curve

This subsection previews Chapter 5 where a more general treatment and full details can be found. There are a few technical details which can be ignored on a first reading.

The basic assumption is that the centre of every galaxy contains a heavy rotating mass (presumably a black hole). It is the ID effects from this mass that cause equatorial orbits to exhibit the characteristic rotation curve.

To fix notation, consider a central mass M at the origin in 3-space which is rotating in the right-hand sense about the z-axis (ie counter-clockwise when viewed from above) with angular velocity ω_0. Assume a flat background space-time, away from M, with sufficient fixed masses at large distances to establish a non-rotating inertial frame near the origin, if the effect of M is ignored. Let P be a point in the equatorial plane (the (x, y)-plane) at distance r from the origin. The rotation of the inertial frame at P is given by adding the contribution from M to the contribution from the distant masses and the nett effect is a rotation of

(3.1) $$\omega(r) = \frac{A}{r + K} \quad \text{where } K = kM \text{ and } A = K\omega_0.$$

The key to the rotation curve is to understand the way in which the inertial drag field affects the dynamics of particles moving near the origin. For simplicity work in the equatorial plane and in a linear approximation to a background Minkowski space. Assume that the inertial frame at P (at distance r from the origin) is rotating with respect to the background with angular velocity $\omega(r)$ counter-clockwise. When computing rotation curves, the formula for $\omega(r)$ just found (3.1) will be used but for the present discussion it is just as easy to assume a general function. The inertial frame at P can be identified with the background space, but it is important to remember that it is rotating. There is no sensible meaning to the centre of rotation for an inertial frame. Two rotations which have the same angular velocity but different centres differ by a uniform linear motion and inertial frames are only defined up to uniform linear motion. Thus it can be assumed for simplicity that all the rotations have centre at the origin. Then the inertial frames can be pictured as layered transparent sheets, each comprising the same point-set but with each one rotating with a different angular velocity about the origin. Each sheet corresponds to a particular value of r. It is necessary to be very clear about the nature of motion in one of these frames. A particle moving with a frame (ie one stationary in that frame) has no *inertial velocity* and its velocity is called *rotational*. In general if a particle has velocity **v** (measured in the background space) then

$$\mathbf{v} = \mathbf{v}_{\text{rot}} + \mathbf{v}_{\text{inert}}$$

where its *rotational velocity* $\mathbf{v}_{\text{rot}}$ is the velocity due to rotation of the local inertial frame and $\mathbf{v}_{\text{inert}}$ is its *inertial velocity* which is the same as its velocity measured *in* the local inertial frame. Note that $\mathbf{v}_{\text{rot}} = r\omega(r)$ directed along the tangent.

Inertial velocity correlates with the usual Newtonian concepts of centrifugal force and conservation of angular momentum.

The fundamental relation

As a particle moves in the equatorial plane it moves between the sheets so that a rotation about the origin which is rotational in one sheet becomes partly inertial in a nearby sheet. For definiteness, suppose that $\omega(r)$ is a decreasing function of r and consider a particle moving away from the origin and at the same time rotating counter-clockwise about the origin. The particle will appear to be being rotated by the sheet that it is in and this causes a tangential acceleration. This acceleration is called the *slingshot effect* because of the analogy with the familiar effect of releasing an object swinging on a string. But at the same time the particle is moving to a sheet where the rotation due to inertial drag is decreased and hence part of the tangential velocity becomes inertial and is affected by conservation of angular momentum which tends to decrease the angular velocity. These two effects balance each other out in the limit and this explains the flat asymptotic behaviour. More precisely, let v be the tangential velocity of the particle in the direction of ω then the slingshot effect causes an acceleration $dv/dr = \omega(r)$ but conservation of angular momentum that operates on the inertial part of v, namely $v - r\omega(r)$ causes a deceleration in v of $(v - r\omega(r))/r$ or an acceleration $dv/dr = \omega(r) - v/r$ and adding the effects we have the *fundamental relation* between v and r:

(3.2)
$$\boxed{\frac{dv}{dr} = 2\omega - \frac{v}{r}}$$

Given ω as a function of r, (3.2) can be solved to give v as a function of r.

(3.3)
$$v = \frac{1}{r}\left(\int 2\omega r\,dr + \text{const}\right).$$

Of interest here are solutions which, like observed rotation curves, are asymptotically constant, and, inspecting (3.3), this happens precisely when $\int 2\omega r\,dr$ is asymptotically equal to Qr for some constant Q. But this happens precisely when 2ω is asymptotically equal to Q/r. This proves the following result.

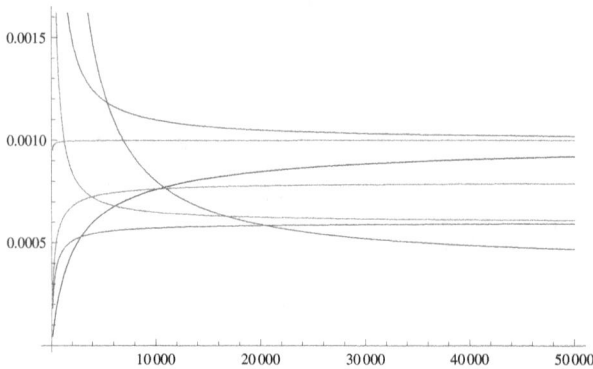

Figure 3.2: A selection of rotation curves from the model

Theorem   *The equatorial geodesics have tangential velocity asymptotically equal to constant Q if and only if ω is asymptotically equal to A/r where $Q = 2A$.*

Equations 3.1 and 3.3 can be combined to give a formula for rotation curves and a selection of the resulting curves is given in Figure 3.2 which should be compared with Figure 3.1 (right).

In passing, it is worth mentioning that in addition to the dark matter problem there is another "problem" often mentioned, namely "dark energy". The author's view is that there is no problem here. Dark energy is just a name for the term in the field equations involving Einstein's cosmological constant, which provides global curvature. It is not a problem any more than the curvature of the earth's surface is a problem: it is just part of the description of the universe!

Spiral structure

The model can be developed to explain the spiral structure. For details and for a Mathematica program to allow the reader to investigate the model, see Chapter 7. For typical output see Figure 3.3.

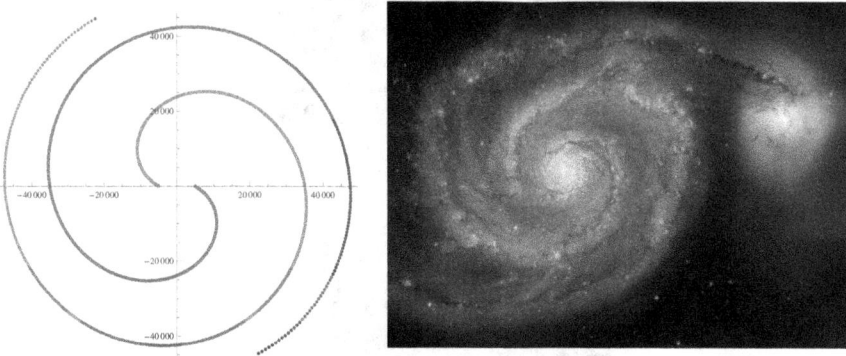

Figure 3.3: Output from the Mathematica program (left) and galaxy M51 from the Hubble site for comparison (right)

3.4 The Arp problem

Halton Arp was a talented observer who provided key observations supporting the HLSW theory of the expanding universe[5]. He also observed many examples of quasars with intrinsic (gravitational) redshift against the current mainstream dictat that all redshift must be cosmological. A particularly striking example is reproduced in Figure 3.4. It is commonsense that the alignments seen in this configuration of galaxies and quasars are not due to chance and there are many similar such in Arp and other's observations [31, 45]. Objects 2 and 3 (quasars) have cosmological redshift around $z = 0.030$ (for the filament) and the remainder must be instrinsic (presumably gravitational). As often happens when a consensus view is challenged by direct evidence, the evidence is ignored and the challenger

---

[5]See the convention on page 21.

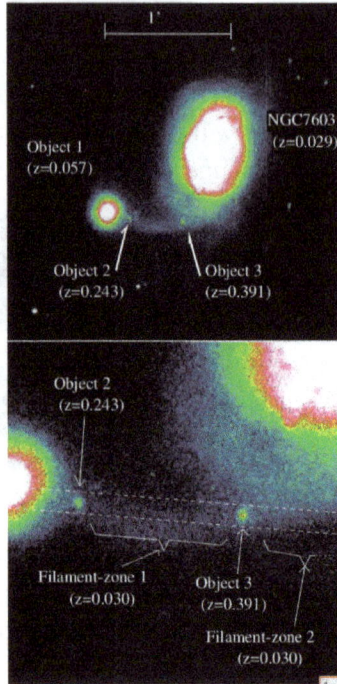

Figure 3.4: NGC 7603 and the surrounding field. R-filter, taken on the 2.5 m Nordic Optical Telescope (La Palma, Spain). Reproduction of Figure 1 of [68]

discredited. Arp was sidelined by the mainstream cosmological community and denied observation time on the big telescopes. One of the main aims of this book is to rehabilitate Arp's reputation (unfortunately posthumous), and if it succeeds in this aim it will have been worth writing. As will be seen shortly, it is in fact quite easy to correct the standard model for quasars to allow for instrinsic redshift, indeed, once again, the correction is to use a simple piece of relativistic geometry.

3.5   The quasar–galaxy spectrum

A major theme of this work is that there is a spectrum of related phenomena. This is the quasar–galaxy spectrum. The unifying element is the presence of

a massive (or hypermassive) black hole. The position of a quasar or galaxy
on the spectrum is determined entirely by the size of this associated black
hole, which varies from 10^6 solar masses (sm), or less, for a small quasar
such as Sagittarius A^*, through 10^9 to 10^{11} sm for a so-called active galaxy
and up to 10^{14} sm, for a full size mature spiral galaxy. An aside here: the
phrase "so-called" for active galaxies is used because one of the main theses
of this work is that all galaxies are highly active and that, for spiral galaxies,
this activity manifests itself in the very spiral structure that characterises
them.

These phenomena are systematically misunderstood in the consensus theory.
At the smaller end, the quasar end, there is an observed redshift which can
be very large (up to $z = 8$ or more — much more as will be seen later)
and, for reasons which will be explained shortly, the current mainstream
view is that this redshift is entirely cosmological (due to the expansion of
the universe). This implies that these objects are very distant, extremely
massive, created just after the Big Bang and have a truly phenomenal power
output, which is very hard to explain. One of the major tasks of this work is
to explain how this view has arisen and how it can be changed to the view
that, by contrast, quasars are typically small, nearby objects with a modest
power output easily modelled by a simple spherical accretion mechanism.

The key to this misunderstanding and to the correct model for quasar energy
production is angular momentum. Very early in the study of quasars it
was decided that the behaviour of angular momentum gives a compelling
reasons for believing that quasar redshift is cosmological. Quasars are
typically believed to be based around a very dense object, probably a black
hole, and their energy production is believed to be due to accretion from
the surrounding medium. Particles fall into the gravitational well of the
central mass and the gravitational energy is released by interaction between
different infalling particles. Now given a small but very heavy object, a
particle approaching with a small tangential velocity will have its tangential
velocity magnified by conservation of angular momentum and there will
be a radius of closest approach. It is very unlikely to actually fall into the
central gravitational well.

The same thing happens for the full flow of infalling matter from the surrounding medium which will typically have a nonzero angular momentum around the black hole. This gives an obstruction to accretion which was found not long after quasars were discovered, for example Michel [75, Section 4, page 158] (1976) states:

> ... One must, however, somehow transfer away most of the angular momentum that the infalling gas had relative to the centre of mass. It seems physically plausible that the effect of such angular momentum would be to choke down the inflow rates. For example, even when magnetic torques are included ... one finds that the 'infall' solutions terminate at finite distances from the origin in analogy with the minimum approach distance of a single particle trajectory having non-zero initial angular momentum. ...

These considerations have led to the subject being dominated by the theory of "accretion discs". The idea is that, since infalling matter cannot flow smoothly into the central black hole, it must typically settle into a rotating structure of some kind, which is called (whatever its actual shape) an accretion disc. Then interaction between infalling matter and this structure allows energy to be produced.

A consequence of this is that redshift, which is frequently observed in quasar radiation, is generally believed to be cosmological and not gravitational (or intrinsic). Indeed if the observed radiation comes from an accretion disc and the redshift is caused by the gravitational field of the nearby black hole, then because the disc varies in its distance from the black hole over its extent, the spectral lines observed would be wide (a phenomenon known as "redshift gradient") and not the narrow lines that are observed.

This in turn implies that the universe has varied in its constitution over the observable past. As remarked above, a quasar with a large cosmological redshift must be a massive object with a huge energy output. But there are no observations of huge sources of energy close to us like these (supposed) near the Big Bang. This provides strong supporting evidence for the Big Bang theory, which entails a continuous change in the constitution of the universe. A steady state model cannot contain a Big Bang.

It was considerations like these that caused Fred Hoyle to abandon his continuous creation model which is fully Copernican in both space and time, ie with no observable global change over time.

3.6   Killing the angular momentum obstruction

One of the main theses of this book is that the angular momentum obstruction to accretion *can be killed by the black hole itself* and this implies that quasars can be relatively small, nearby objects and the universe could be Copernican in time as well as space. Thus Hoyle's model could still be correct (though it is not the model proposed here).

The key to killing this angular momentum obstruction is to work in a relativistic framework and not the Newtonian framework implicitly assumed in the above discussion. The relativistic effect of "inertial drag" (ID) used for the rotation curve can also be used to explain how a black hole can compensate for the angular momentum of the infalling gas/plasma stream. This allows for an energy production model to be established with radiation coming from a thin spherical region (the Eddington sphere) which can be very close to the event horizon of the black hole and subject to an arbitrarily high gravitational redshift, with the cosmological redshift small in comparison. Because the production sphere is thin, there is little redshift gradient. For details here, see Section 6.1.

As mentioned above there is some very strong evidence in the observations of Arp and others [31, 45] that quasars do in fact possess intrinsic, ie gravitational, redshift. This and the angular momentum considerations just mentioned led Arp to propose some fantasy physics explanations for this redshift. The explanation proposed in this work uses only well-accepted (and definitely not fantasy) physics and is fully consistent with Arp's observations.[6]

---

[6]The one new hypothesis that is made in this work, the inertial drag force, see below, does not play any role in this explanation.

3.7   Embedding Mach's principle in EGR

Alongside inertial drag, the other main ingredient for the new model presented in this book is "Mach's principle" outlined above. It is necessary to explain how to embed the version of Mach's principle that is needed for this work into EGR (Einstein's General Relativity). There are obvious causal problems in a naive statement: how exactly does distant matter communicate with local matter to determine the local inertial frame? and does the influence happen instantaneously or travel at the speed of light? These problems will be avoided by restricting to a limited version of the principle due to Sciama [89] which is quantitative rather than philosophical and which is referred to as Sciama's principle. The discussion is further simplified by concentrating on rotation at the expense of general non-inertial motions. EGR deals well with acceleration, so this makes sense for the purposes of embedding the principle within EGR.

A new hypothesis is needed for the dragging effect of a rotating body on the inertial frames near it. The precise behaviour that is needed is not a consequence of Einstein's equations and the hypothesis amounts to assuming that a rotating mass has a non-zero effect on the stress-energy tensor near it — in other words stops the space near it being a true vacuum. This gives a natural way to understand how inertial drag propagates: the disturbance to the local vacuum is akin to a gravity wave and propagates at the speed of light. Furthermore reading back from the rest of the universe, the local background inertial frame is created by the rest of the universe by a similar propagation effect from all the rest of the matter (a brief aside here: this makes sense only if the sum is finite — or quasi-finite — this will be explained in the next chapter).

It is worth briefly comparing the new inertial drag hypothesis made in this book with the dark matter hypothesis made in current mainstream cosmology. At first sight they may appear to be similar. Both correct the rotation curve for galaxies. But the dark matter hypothesis amounts to assuming the existence of inert matter, which has no effect other than gravitational attraction, and cannot otherwise be detected. The inertial drag hypothesis on the other hand amounts to assuming a new effect of a rotating body on the field outside it. It embodies a limited version of Mach's principle which, as has been seen, is philosophically compelling, and must be embodied in any theory that seeks to accurately describe reality. Thus, unlike the dark matter hypothesis, the inertial drag hypothesis is a necessary part of a complete theory. Furthermore, the inertial drag hypothesis also underlies a good model for the dynamics of spiral galaxies, whereas the dark matter hypothesis leaves this problem unsolved. More detail on this point will be given later in the book (Section 4.8).

3.8 Outline of Part 2

This is the last section in Part 1. But do not despair. All the chapters in Part 2 contain explanations accessible to a general reader as well as technical details. Read on; omit the more technical parts; they may well be accessible on a second or later reading. Here is an outline of the ideas covered in Part 2.

Mach's principle is discussed in Chapter 4, after which Chapter 5 derives the inertial drag effect, that allows quasars to cancel out the angular momentum obstruction to accretion, and fuels the dynamics of galaxies. In this chapter it is applied to model the rotation curve for galaxies without needing "dark matter" (as outlined above).

Next in Chapter 6 the subject of quasars is taken up in earnest. Here it is explained how inertial drag allows black holes to absorb the angular momentum in infalling gas / plasma and to grow by accretion. The spherical accretion model that this allows is joint work with Rosemberg Toala Enriques and Robert MacKay [88]. This work is still in draft form, but nevertheless the model fits observations extremely well, including those of Arp [31], and also explains the apparently paradoxical results of Hawkins [52]. This section contains a first description of the pivotal quasar–galaxy spectrum. Technical details from [88] are deferred to Appendix C.

After this the second main task of the book is tackled in Chapter 7, namely to provide a model for the spiral structure of full-size galaxies, such as the Milky Way, which lie at the other end of the quasar–galaxy spectrum.

The nature of these objects is also much misunderstood by mainstream cosmology. Spiral galaxies all contain a central hypermassive black hole (of mass 10^{11} sm or more), which controls the dynamic by the same inertial drag effect that allows accretion in quasars, and which is surrounded by an accretion structure responsible for generating the visible spiral arms. Another aside here: there is a special misunderstanding with the Milky Way, where SgrA* with a mass of only 4.3×10^6 sm, far too light to have any dynamic effect on the galaxy, is believed to be the central black hole. This misunderstanding will be cleared up at a later stage.

Between quasars and spiral galaxies lie "active" galaxies for which accretion structures have been directly observed. This is the only part of the quasar–galaxy spectrum which is more-or-less correctly understood by mainstream cosmology. There will be a lot more to say about the whole quasar–galaxy spectrum later in this work.

As mentioned above, there is, inside a full size spiral galaxy, an accretion structure, called "the generator", which is responsible for generating the spiral arms. This is described in Chapter 7 where a full model for the resulting spiral structure is derived. The generator feeds the roots of the spiral arms with a pure light element mixture (H and He with a trace of Li). This is the same mixture of elements that is hypothesised to have been created in the Big Bang just before the time of the last scattering surface from the cooling of a hot plasma of quarks, and the process is similar. The residue of these streams, not condensed into stars, escapes the galaxy and feeds the intergalactic medium and this explains the observed proportion of these elements in the universe (which is one of the so-called "pillars of the Big Bang theory").

Chapter 8 and Chapter 9 cover observations and consequences for cosmology. Included here are a comprehensive rebuttal of the Big Bang theory and explanations for redshift and the Cosmic Microwave Background (CMB), which are the other two pillars. Both are based on the proposed de Sitter space model for the universe, which also provides an explanation for Gamma Ray Bursts (GRB). Technical details for several of the topics are again deferred to appendices.

PART 2

A more technical treatment

Chapter 4

Sciama's principle

This chapter is concerned with a discussion of Mach's principle and the restricted version that is needed for the dynamical applications (to quasars and spiral galaxies) in the rest of the book. The final form of the principle (the Weak Sciama Principle) hypothesises an inertial dragging effect from a rotating body which drops off asymptotically with k/r where k is a constant and r is distance from the centre. A reader who is happy to accept this principle can omit this chapter without loss. The precise assumption is repeated near the beginning of the next chapter.

4.1   Inertial frames and Mach's principle

In any dynamical theory there are certain privileged frames of reference in which the laws of Newtonian physics hold to first order. These frames are variously called "inertial frames" or "rest frames". They are characterised by a lack of forces correlated with acceleration or rotation. In Newtonian physics there is a universal inertial frame referred to as "absolute space" and in Minkowski space the standard coordinates provide an inertial frame at the origin. Then Lorentz transformations carry this frame to an inertial frame at any other point, providing inertial frames for special relativity. General relativity is built on Minkowski space which in turn provides

47

inertial frames for this theory, see Section A.6. Berkeley [32] and Mach [69] criticised Newton's assumption of absolute space. Berkeley suggested that the local rest frame could be defined by distant "fixed" stars. Mach's book [69, Ch II.VI (page 271 ff)] contains a devastating critique of Newton's assumptions and is well worth reading. It was extremely influential and Einstein acknowledged a debt to his ideas. Mach's basic point is that one should never assume anything that is not directly connected to observations of some kind and in particular the concept of the local inertial frame must be defined in terms of (theoretically) observable quantities. Some detail from Mach is given in Section 4.3 below.

The basic property of inertial frames is that they are only defined up to uniform linear motion. Given any inertial frame, a frame which is in uniform linear motion with respect to the given frame is also an inertial frame. Thus "the" inertial frame at a point P in fact means an equivalence class of frames, two frames in the class being in mutual uniform linear motion with respect to each other. (For this reason, calling them "rest" frames is highly misleading and this terminology will not be used again.)

Mach's ideas have passed into general circulation as "Mach's principle" which is usually summarised as stating that the local concept of inertial frame is correlated with the distribution and motion of all the matter in the universe. However there are many other ways of interpreting the principle and there is a huge literature on the subject. At its weakest, the principle is interpreted as merely stating that all phenomena must have their origin in some material source (see eg [90]), and it has even been interpeted as an assumption about the nature of the Big Bang (Tod [99]).

For the purposes of this book, a statement is needed which is more precise than these but not so wide ranging. What is needed is a local version which applies to rotation of inertial frames and which is quantified precisely.

4.2   Sciama's principle

The version that is used is close to the version in Sciama's thesis [89]. Sciama makes a bold attempt to base a full theory of dynamics on Mach's principle. His idea is that the inertial frame at any point P in the universe is determined by the inertial frames at every other point Q. The contribution from Q is nonzero only if there is a mass m_Q at Q and then the contribution is (a) proportional to this mass and (b) inversely proportional to the distance r_Q between P and Q. In other words the contribution is

$$m_Q \, \text{IF}_Q / r_Q,$$

where IF_Q means the inertial frame at Q. The idea is that this should be summed over "all the matter in the universe".

To make sense of this sum it is necessary to make a number of assumptions. Firstly, in order to add up contributions, it is necessary to work in a linear framework and the simplest way to do this is to work with a perturbation of flat (Minkowski) space, which is exactly what Sciama does. The underlying Minkowski space provides "standard" reference frames at each point and the motion of any frame can be measured with respect to this standard, and also provides a space in which to measure the distance r_Q used in the summation.

Working within a perturbation of Minkowski space limits the theory to weak fields, but it suffices for most of this work. When working near the massive centre of a galaxy, use can be made instead of a perturbation of any spherically-symmetric metric, eg the Schwarzschild metric, which allows stronger fields.

Secondly, in order for the summation to converge, the "universe" needs either to be finite or to be "quasi-finite" in the sense that only a finite part contributes to the sum. More detail on this point is given below.

Finally, it is necessary to keep r_Q from getting too small or else the contribution of m_Q will be far too large. This can be done either by ignoring masses which are close to P, since the factor $1/r$ implies that the sum is dominated by distant matter, see the discussion below, or, if there is a significant and very massive body (eg the black hole at the centre of a galaxy) nearby, then the sum can be normalised as explained below.

To formulate the principle quantitatively use the notation NM_P for the the non-uniform motion of the inertial frame at P and ditto Q, in other words its acceleration and/or rotation measured with respect to to the local reference frame, then the inertial frame at P is given by the reference frame plus NM_P and the principle states that

Sciama's principle $$NM_P = K \sum_Q \frac{m_Q}{r_Q}(NM_Q).$$

This statement is digested from Sciama's introduction and the precise formulation in terms of the field [89, Equation (1), page 37]. It is called *Sciama's principle* in order to distinguish it from Mach's principle. Here K is a normalising factor which will be dicussed further below.

Notice that this principle is completely symmetric. The effect of Q's motion on the inertial frame (IF) at P is exactly similar to the effect of P's on the frame at Q. And note that the effect is *coherent* in the sense that an acceleration or rotation of the frame at Q causes an acceleration or rotation of the frame at P with the *same* direction or sense. Sciama describes this symmetry eloquently in his introduction, for example: "... *the statement that the Earth is rotating and the rest of the universe is at rest should lead to the same dynamical consequences as the statement that the universe is rotating and the Earth is at rest, ...* "

Also notice that using NM_P in the summation implies that the inertial effect of matter in uniform linear motion is ignored. This is correct for small masses or for larger masses sufficiently distant that gravitational induction effects can be ignored.

With a caveat that this needs needs to be treated with care in special cases, this will be adopted as a working hypothesis which fits the intuitive idea of inertial effects:

Working hypothesis *Uniform linear motion has no inertial effect.*

Sciama is clear that his principle is incompatible with Einstein's General Relativity (EGR) and is attempting to create an alternative theory. Later it will be seen precisely how the principle is incompatible with EGR and it will be explained how to modify EGR to include the principle for rotation (by interpeting the principle as adding a stress field that causes the inertial drag and radiates from the rotating mass).

Sciama starts to derive a full gravitational theory from this principle. He specialises to a "field" (a vector field) defined on Minkowski space and as he makes clear this is an interim approach which will need improvement is a subsequent promised sequel paper (which in fact was never written). In order for the summation to converge, the "universe" needs either to be finite or to be "quasi-finite" in the sense that only a finite part contributes to the sum. More detail on this point is given in the next paragraph.

Sciama discusses three cases in detail:

(a) The effect of distant matter on the local IF.

The factor $1/r$ is chosen to make distant matter dominate. In order to get a finite sum, Sciama assumes standard HLSW expansion (see the convention on page 21) and then it is natural to limit the summation to the visible universe (in other words to ignore parts that are regressing faster than c). It is worth remarking in passing, that it is not necessary to assume the existence of a Big Bang (BB) to satisfy this quasi-finite hypothesis. There are models for the universe with redshift fitting observations but with no BB (cf Section 9.4), the simplest of which is the expanding part of de Sitter space; there is also the (now largely ignored) continuous creation model of Hoyle et al [54]. The effect of distant matter needs to be normalised to unity. For example, if the whole universe is rotating about P with angular velocity w, then this should induce a rotation of w in the IF at P, in other words the situation should be exactly the same as if all were at rest. Similarly for acceleration. Thus

(4.1)
$$K \sum_Q \frac{m_Q}{r_Q} = 1$$

where the sum is taken over all accessible matter Q (ie within the visible universe). One way to arrange this is to assume that $K = 1$ and

(4.2)
$$\sum_Q \frac{m_Q}{r_Q} = 1$$

This makes perfect sense provided that r_Q is never small (if r_Q is allowed to tend to zero, the contribution from m_Q goes to infinity, which is absurd) and this is effectively what Sciama does. A more sensible way is to normalise by setting

(4.3)
$$K = 1 / \sum_Q \frac{m_Q}{r_Q}$$

which compensates for large local masses and this is what will be done when the principle is applied near the large central mass of a galaxy.

Equation (4.2) implies a fundamental relation between the various gravitational and cosmological constants which Sciama derives as [89, Equation (7)]. He points out that this is, within reasonable limits, in accord with observations. Misner, Thorne and Wheeler (MTW) [76, below 21.160] make exactly the same point using more modern observations[1]. This provides a preliminary justification for the key factor $1/r$. A better justification comes with the simple model Sciama describes, where the field naturally decays like $1/r$. His model however is too simplistic (as he readily acknowledges) and in fact coincides with one of the standard approximations to EGR, namely "gravitomagnetism". Shortly, there will be other cogent reasons for the factor $1/r$.

For the other two cases he uses the model.

(b) A locally isolated mass. Here Sciama finds Newtonian attraction to first order (and in fact it is always attraction).

(c) A locally rotating frame. Here he finds the usual Newtonian story (Coriolis forces etc).

4.3   An excerpt from Mach's critique

There is a passage in Mach's critique which can be used to provide further support for the factor $1/r$. In Ch II.VI.7 (page 286) of [69] he points out that if two bodies move uniformly in ordinary 3-space then each sees the other as having non-zero acceleration along the common line of sight. Uniform motion does not *appear* uniform. Indeed if r is the distance between the bodies, then

(4.4)
$$\frac{d^2r}{dt^2} = \frac{1}{r}(u^2 - v^2)$$

[1]There are about 10^{11} galaxies in the visible universe of weight about 0.03 (10^{11} solar masses) at distances varying up to 10^{10}, where natural units are used (G (Newton's gravitatonal constant) $= c = 1$ and everything is in measured in years).

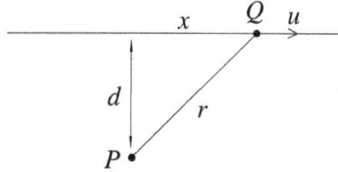

Figure 4.1: Proof of Mach's formula for apparent acceleration of bodies in uniform relative motion. Here $dx/dt = u$ (const) and $dr/dt = v$. Differentiating $r^2 = d^2 + x^2$ twice, gives $r\,d^2r/dt^2 + v^2 = u^2$ and hence $d^2r/dt^2 = (1/r)(u^2 - v^2)$ with $|v| = |(x/r)\,u| < |u|$.

where u is the absolute value of the relative velocity and $v = dr/dt$. This is readily proved from Pythagoras' Theorem, see Figure 4.1. And notice that $|v| < |u|$. Thus to describe even uniform motion in terms of observation is quite complicated. On the next page he gives a formula for the mean acceleration of a body P with respect to a system of other masses (weighted by their mass) namely

(4.5)
$$\sum m_Q \frac{d^2 r_Q}{dt^2} \Big/ \sum m_Q$$

where m_Q is at distance r_Q from P. The notation (but not the formula) has been changed in order to show the connection of Mach's analysis with Sciama's principle. Because now, if it is assumed that all bodies move uniformly with bounded mutual velocity and if (4.4) is substituted in (4.5), the following formula for the acceleration of P in terms of the other masses is found

$$\sum \frac{m_Q}{r_Q} b_Q$$

where $b_Q = (u_Q^2 - v_Q^2)/\sum m_Q$ are all bounded by $1/\sum m_Q$ times the square of the bound for the mutual velocities. This is very close to Sciama's principle (ignoring rotation). To see the connection, drop the assumption of an absolute space where all this was supposed to take place. Keep only the observations. This equation can be interpreted as specifying the "absolute" acceleration of P (and hence the IF at P) in terms of data at Q and if these data are labelled "inertial effect" then this would obtain precisely Sciama's principle.

It is important to remark that this discussion is not intended to suggest that uniform motion has an inertial effect; a small mass moving uniformly has negligible inertial effect, though a large mass has some effect due to inductive effects from its gravitational field. What is intended is that the formula that it is sensible to use to estimate the local inertial frame is likely to include a factor $1/r$ since apparent acceleration due to uniform motion does indeed include such a factor. The discussion is intended to support the contention that inertial effects drop off like $1/r$.

4.4   Rotation

Non-inertial motions are combinations of acceleration and rotation (and inertial motions). Now EGR deals well with acceleration. This is in some sense its major application, but as will be seen, it does not deal well with rotation. So for the purposes of embedding Sciama's principle in EGR, it makes sense to concentrate on rotation.

Sciama's principle applied to rotation says that rotation of a mass m_Q at Q contributes $K m_Q \omega_Q / r_Q$ to the rotation of the IF at P where ω_Q is the angular velocity of m_Q.

It is important to notice that it is the angular *velocity* of m_Q which contributes to the sum and not the angular *momentum* of m_Q about P. This behaviour (and a further final argument supporting the key factor $1/r$) can be deduced from a simple dimensional agument. There is a highly relevant passage in MTW [76] discussing precession of the Foucault pendulum which is worth quoting extensively. It starts on page 547 para 3 with the margin note *The dragging of the inertial frame*. It has been edited very slightly to make the notation fit with the present discussion and to suppress mention of conventional units. In this book *natural units*, with $G = c = 1$ and everything measured in years are used for most of the calculations; here G is Newton's gravitational constant and not Einstein's tensor which is also commonly denoted G.

> *Enlarge the question. By the democratic principle that equal masses are created equal, the mass of the earth must come into the bookkeep-ing of the Foucault pendulum. Its plane of rotation must be dragged around with a slight angular velocity, ω_{drag}, relative to the so-called*

> *"fixed stars." How much is ω_{drag}? And how much would ω_{drag} be if the pendulum were surrounded by a rapidly spinning spherical shell of mass m_{shell} and radius r_{shell} turning at angular velocity ω_{shell}? Einstein's theory says that inertia is a manifestation of the geometry of space-time. It also says that geometry is affected by the presence of matter to an extent proportional to the factor G/c^2 (ie 1 in natural units). Simple dimensional considerations leave no room except to say that the rate of drag is proportional to an expression of the form*

$$\text{(21.155)} \qquad \omega_{drag} = k \frac{m_{shell}}{r_{shell}} \omega_{shell}.$$

> *Here k is a factor to be found only by detailed calculation. ...*

Details of the dimensional argument used here will be given later. The authors continue by discussing the results of Lense and Thirring where k is calculated to be $4/3$ assuming a specific approximation which is in fact identical to the Sciama field. There will be more to say about this shortly.

At this point it is worth making an observation. The Sciama field can be seen as a first approximation to a full-blown theory of dynamics based on Mach's principle. Since it coincides with gravitomagnetism, which is a first approximation to EGR, it follows that no local observations, where the fields are weak (for example the Gravity Probe B experiment [42]) can distinguish between EGR and a theory of dynamics based on Mach's principle. One of the main theses of this book is that there is however strong experimental evidence in favour of the latter from observations of galaxies.

4.5   The weak Sciama principle

Continuing the discussion of the MTW quotation and equation (21.155), their "democratic principle" is close to Sciama's principle, at least in its universality, referring as it does to all (accessible) matter in the universe. The equation itself is precisely the principle for the contribution of the mass m_{shell}. And notice that it is implied that the dragging effect of the earth should be coherent with the earth's rotation. This point is so obvious that it may easily be overlooked and is only mentioned because shortly a model will be examined where the dragging is not always coherent. To see the connection with Sciama's principle for many distinct rotating masses, consider the following thought experiments. Replace the shell by a ring of matter at distance $r = r_{shell}$. Nothing changes qualitatively. The constant k reflects the precise geometry of the setup and may change. Now imagine that the ring is a necklace of n beads all of the same mass m. By the democratic principle, each has the same effect $\omega'_{drag} = \omega_{drag}/n$ and, if P is the centre of the ring and Q one of the beads, then Q contributes $km\omega/r$ to the inertial frame at P where ω is the angular velocity of Q moving around P. But the local motion of Q is exactly the same as a (uniform) linear motion of velocity ωr along the tangent together with a rotation on the spot of ω. Using the working hypothesis, the linear motion has no inertial effect and the formula for the drag is now exactly Sciama's principle in this case, namely:

Weak Sciama Principle *A mass m at distance r from P rotating with angular velocity ω contributes a rotation of $km\omega/r$ to the inertial frame at P where k is constant.*

This *weak Sciama principle* is the statement that is needed for the dynamics of galaxies. The constant k is a normalising factor which needs to be set in context. When the principle is used in the next chapter (Equation 5.1), this will be made precise.

Incidentally it can now be seen why the working hypothesis implies that angular momentum is the wrong measure of the inertial effect of one mass on another. A uniform linear motion has no inertial effect, but, adding a linear motion to Q may well have a strong effect on its angular momentum about P. Conversely rotation need not correlate with angular momentum: If Q is in fact a point mass, then rotation of Q with angular velocity ω has no angular momentum about P whereas motion in a circle around P with the same angular velocity does have angular momentum.

The weak Sciama principle is not Machian in even the weakest version (that all effects are due to observable source). It makes no attempt to completely specify the IF at P in terms of all the matter in the universe and indeed it leaves open the possibility that the IF at P may be affected by unknown events (perhaps they are outside the visible horizon — cf MacKay–Rourke [72] and Appendix F). But the advantage of a local statement of this type is that it avoids the causality problems implicit in any global statement and it is open to direct verification using local observations. One of the main theses of this book is that it is indeed strongly supported by observations of galaxies and in particular their characteristic rotation curves.

4.6 The Lense–Thirring effect

Like the full Sciama principle, the weak principle only makes sense in an approximation to Minkowski space and this is exactly how it will be used (the formulation is given near the start of the next chapter). Early work of Lense and Thirring [97] mentioned above, calculated the inertial drag due to a heavy rotating body assuming a specific approximation to EGR. To be precise they calculated the inertial drag due to a rotating spherical shell for points nearby. As seen above, this effect is roughly in accord with Sciama's principle for points inside the shell, but as will be seen shortly, it is hopelessly wrong outside.

The approximation they used is the same as that used by Sciama and is known as gravitomagnetism. The equations correspond formally to Maxwell's equations and the effect can be understood by thinking of electromagnetism. Motion of matter corresponds to electrical current and a circular motion induces a linear magnetic effect. The dragging effect corresponds to magnetic lines of force with the induced rotation having the line as axis with rotation around the line in the positive sense. Thus a rotating body behaves like a magnet and causes inertial drag which is coherent near the poles but anti-coherent to the side where the magnetic lines run back between the poles.

This has some very counter-intuitive consequences.

(a) Uniform linear motion has rotational inertial effects.

(b) A rotating body drags some frames nearby in the opposite direction to the rotation causing the drag.

(c) In general the direction of drag is unrelated to the rotation which induces it.

Effect (b) was picked up by Rindler [84] and correctly labelled "anti-Machian". However his conclusion that Mach's principle needs to be treated with care "one simply cannot trust Mach!" is bizarre. The philosophical reasons for Mach's principle are compelling and it must be incorporated in any theory that describes reality. It is the Lense–Thirring effect that must be wrong. In any case, it is not necessary to appeal to Mach's principle to see that inertial drag should be coherent. As will be seen in a couple of lines, a simple thought experiment using general principles of symmetry and continuity will establish this fact.

4.7   Central rotation

Perhaps the Lense–Thirring effect is wrong because of the approximation used, so now turn to theories without approximation, including EGR. Consider a dynamical theory, which may not be EGR, but which is metrically based and which specialises to special relativity locally in same way that EGR does, with a similar equivalence principle. Here is a simple thought experiment which shows that, in any such theory, frame dragging due to a central rotating body exists and is coherent.

Imagine that the universe is a 3-sphere (spatially) and that it is filled with two very heavy bodies (both 3-balls) with a comparatively small (vacuum) gap between them. Suppose that these bodies are in relative rotation. Then by symmetry, frames half way between the bodies will rotate at the average speed and by continuity the inertial drag will move towards rotation with each of the bodies as one moves away from the centre. Diagramatically the situation is pictured in Figure 4.2. Note that in the figure the bodies are represented as nested. To get the correct view think of the outer circle labelled "infinity" as the diametrically opposite point to the centre of the inner body. To make sense of inertial drag here, assume that the space between the two bodies has a flat background metric, but do not assume anything about the space inside the bodies.

Now shrink the inner body to be the central rotating body and imagine the outer body to be the rest of the heavy universe. It is unreasonable to

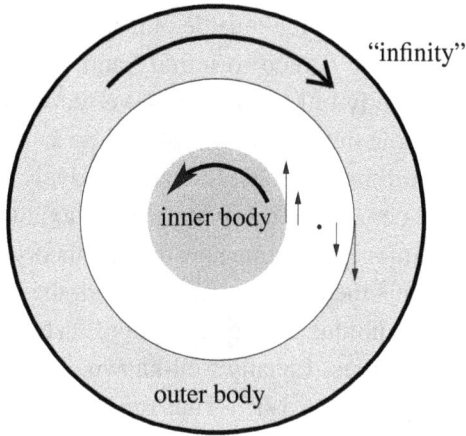

Figure 4.2: Inertial drag between two heavy bodies

suppose that the qualitative description of inertial drag changes during the shrinking process and therefore, in the final metric, the central body will induce coherent inertial drag.

Now appeal to the same dimensional considerations as used in the passage quoted from MTW (above) to deduce the weak Sciama principle. Let the rotating body be labelled Q and have mass m and angular velocity ω. For simplicity, consider a point P on the equatorial plane of the rotating body at distance r from the centre. It is a commonsense assumption that the dragging effect at P is proportional to $m\omega$ and, being a pure rotation of the local inertial frame, has dimension $1/T$ where T means "time". (Notice that there is no sensible meaning to the centre of rotation for this effect. Two rotations which have the same angular velocity but different centres differ by a uniform linear motion and inertial frames are only defined up to uniform linear motion.) Now in relativity time, mass and distance all have the same dimension. Thus $m\omega$ is dimensionless and the only sensible formula for the induced inertial drag is $km\omega/r$, possibly normalised (note in passing that normalising constants such as k are dimensionless and do not affect this argument).

Go further with this thought experiment. Assume now that the universe is $\mathbb{R}^3$ spatially with the heavy inner rotating body at the origin. And imagine that the outer body is the outside of a sphere of radius R say and is in fact at rest. Now let R tend to infinity and, as it does so, control the mass of the outer body to keep its inertial effect near the inner body constant. In the limit, the outer body is replaced by an asymptotically flat metric near infinity and then, outside the inner body, is a metric which is stationary (the whole construction was stationary) axially-symmetric and asymptotically flat at infinity. Assume now that the theory being considered is in fact EGR. Then this metric must coincide with the Kerr metric which is well-known to be the unique metric satisfying Einstein's equations for a vacuum with these properties. Equally well-known, the inertial drag effects of the Kerr metric drop off like $1/r^3$. Thus this metric, constructed using the thought experiment, continuity and dimensional arguments, does not satisfy Einstein's equations; or rather, to be very precise, if it is assumed that *the space constructed is a vacuum outside the hypothesised masses*, then the metric does not satisfy Einstein's equations.

One may wonder why the dimensional argument does not equally apply to the Kerr metric. This is because the Kerr metric has a well-defined angular momentum but no well-defined angular velocity. Thus the inertial drag effect of the Kerr metric must be proportional to angular momentum NOT to mass times angular velocity. But angular momentum has dimension T^2 and to get a drag effect of dimension $1/T$ a formula of the type kA/r^3 is needed where A is angular momentum.

It is worth at this point recapping why angular momentum is the wrong measure for inertial effects. This is a simple consequence of the working hypothesis that linear motion has no inertial effect. Angular momentum can be altered by adding a linear motion. Angular velocity cannot be so altered.

4.8   Adding Sciama's principle to EGR

At this point the story seems to have run into an impasse. Assuming the universe obeys standard relativity (EGR) then the version of Mach's principle that is needed does not hold. Inertial drag drops off as $1/r^3$ in EGR and not $1/r$.

There are two sensible ways out of this impasse.

(1) The *revolutionary* approach is to abandon EGR and build a new theory which satisfies Sciama's principle.

(2) The *conservative* approach is to continue to use EGR but add a hypothesis within EGR that implies Sciama's principle. As seen above, this is impossible assuming the space between bodies is a vacuum, so this approach entails hypothesing that *space near a rotating body is not a vacuum* and the thought experiment conducted above is impossible because the space between the rotating bodies is not a vacuum.

This book adopts the conservative approach. Apart from avoiding the non-trivial problem of finding a theory to replace EGR, this approach has one great technical advantage: it provides a mechanism for Mach's principle (at least as it applies to rotation) which does not run into causal problems.

The hypothesis added to EGR is that any rotating body disturbs the local space-time by dragging inertial frames near it coherently by an amount proportional to the rotating mass times its angular velocity, with the influence dropping off asymptotically with k/r where r is distance from the centre of gravity of the rotating mass and k is constant. The precise formula is given in the next chapter, where there is also an interpretation in terms of the metric.

In a vacuum, EGR does not have this inertial drag effect. The Kerr metric which is the only rotationally symmetric vacuum metric flat at infinity and valid in EGR has a drag effect dropping off much faster than this (asymptotically with k/r^3). So the hypothesis amounts to assuming that a rotating mass has a non-zero effect on the stress-energy tensor near it — in other words stops the space near it being a true vacuum. It also gives a natural way to understand how inertial drag propagates: the disturbance to the local vacuum is akin to a gravity wave and propagates at the speed of light. Furthermore reading back from the rest of the universe, the local background inertial frame is created by the rest of the universe by a similar propagation effect from all the rest of the matter. Thus the hypothesis gives a natural causal framework for Mach's principle. An example of this causal framework working in practice would be the case where a rotating body undergoes a sudden change (eg breaking up) which changes the inertial drag field that it causes. This makes a disturbance in the local space-time (a sort of gravity wave) which propagates at the speed of light with no causal problems.

Another consequence is that a rotating body interacts directly with surrounding matter and indeed energy can be extracted in a similar way to the Penrose effect which extracts energy from the Kerr metric. This implies that the rotation will eventually radiate away. This is an extremely small effect for ordinary rotating bodies and only becomes significant for rotating black holes where the energy radiating away fuels the surrounding dynamic as will be seen in the next few chapters. The effect of this can be seen graphically in the spiral structure of full-size galaxies, eg the so called Whirlpool galaxy, Figure 7.7, left.

Superficially the change to vacuum that the new hypothesis entails may seem like an alternative formulation of "dark matter" but it is in fact quite different. The dark matter hypothesis amounts to assuming the existence of inert matter, which has no effect other than gravitational attraction, and cannot otherwise be detected. It is an *incident* hypothesis in the sense that it contains nothing more than what is needed to correct the rotation curve; it is a "fudge factor", designed to correct a shortfall. The inertial drag hypothesis on the other hand amounts to assuming a new effect of a rotating body on the field outside it. It is justified by Mach's principle which, as has been seen, is philosophically compelling and must be embodied in any theory that seeks to accurately describe reality. Thus it is a *necessary* hypothesis in the sense that it needs to made, independently of the rotation curve, in order to encode the necessary Mach principle. The inertial drag field that is assumed to exist can be detected directly by its effect on inertial frames so it has an existence independent of the rotation curve that it serendipitously also predicts.

In EGR a rotating mass does in fact have an effect on the field outside the body, but this is confined to the skew-symmetric part of the field, the *Weyl* tensor, or *trace-free* part of the curvature (which suggests that using the Weyl tensor as well as the Ricci tensor might provide a replacement theory for EGR satisfying Sciama's principle, ie the *revolutionary* approach on page 61). The inertial drag hypothesis implies that a rotating body also affects the other part, the *Ricci* curvature. Einstein's equations for a vacuum are equivalent to the vanishing of the Ricci curvature (see Section A.9). Thus, if the Ricci curvature is nonzero, the field is not an Einstein vacuum.

4.9   Sciama's principle and black holes

Applying Sciama's principle to black holes entails assuming that a black hole has a well-defined angular velocity as well as a well-defined angular momentum. Equivalently a black hole has an effective radius, r_{eff}, related to angular momentum Ω and angular velocity ω by

(4.6) $$\Omega = M\omega r_{\text{eff}}^2.$$

For a black hole the fiction is that the actual radius is zero (total gravitational collapse) and hence angular velocity is not determined. So this assumption is equivalent to replacing conventional theory by the more sensible assumption that, in the collapse to a black hole, matter reaches a small but non-zero size.

4.10   Coda

Sciama's initiative, to base a dynamical theory on Mach's principle as formulated in Sciama's principle, has never been followed up and this approach to dynamics remains dormant. One of the aims of this book is to reawaken this approach. Sciama did return to the topic of Mach's principle in [90]. However this paper abandons Sciama's principle and formulates Mach's principle in one of its weakest forms, namely that all phenomena have their origin in some material source or boundary condition. Moreover the theory exposited in [90] is EGR which as has been seen is incompatible with even the weak Sciama principle.

Chapter 5

The rotation curve

The *rotation curve* of a galaxy with an equatorial plane (for example a spiral galaxy has its spiral arms lying roughly in such a plane) is the plot of tangential velocity against distance from the centre for a particle (star or similar) moving in the equatorial plane. In practice it is not possible to observe one star, but rather the general motion of all stars (or other radiating matter) in the equatorial plane. This makes the observed nature of rotation curves all the more striking. Typically the curve (of tangential velocity against distance from the centre) comprises two approximately straight lines with a short transition region. The first line passes through the origin, in other words rotation near the centre has constant angular velocity (plate-like rotation); the second is horizontal, in other words the tangential velocity is asymptotically constant, see Figure 5.4 (right) below. Furthermore, observations show that the horizontal straight line section of the rotation curve extends far outside the limits of the main visible parts of galaxies and the actual velocity is constant within less than an order of magnitude over all galaxies observed (typically between 100 and 300km/s) see Figure 5.5.

Galactic rotation curves are so characteristic (and simple to describe) that there must be some strong structural reason for them. They are very far indeed from the curve obtained with a standard Keplerian model of rotation under any reasonable mass distribution. In a Keplerian model, suppose that the mass within a radius r of the centre is $M(r)$ then equating centrifugal force with gravitational attraction gives

$$\frac{v^2}{r} = \frac{GM(r)}{r^2}$$

where v is tangential velocity and G is Newton's gravitational constant (taken to be 1 in natural units). Thus if v is asymptotically constant then $M(r)$ is asymptotically equal to a constant times r and tends to infinity with r.

Nevertheless, in spite of the huge mass needed, a Keplerian model is exactly what is assumed in current cosmological theory. To square the circle, current theory hypothesises the existence of a huge amount of matter. Since this matter is not observed, it is called called "dark". It needs to be distributed in precisely the right way to make Keplerian rotation fit the rotation curve. This is extremely implausible for several reasons. Firstly it has just been seen that the quantity of dark matter required is huge and tends to infinity with the radius of fit, which as mentioned above appears to be unbounded. Secondly it is unreasonable to suppose that exactly the right distribution of dark matter happened (by condensation) for every galaxy and thirdly, the final arrangement with most of the matter on the outside is dynamically unstable. For stability in a rotating system (such as the solar system or Saturn's discs) there must be a strong central mass to hold it together. Failing this the system will tend to condense into smaller systems. Finally despite the best efforts expended in the search, nor hair nor hide of dark matter has been found to date.

This chapter presents a solution to these problems using a quite different point of view. The suggestion made here is that the centre of a typical galaxy contains a huge rotating body (probably a black hole) and that the inertial drag effects coming from this rotating mass are responsible for the observed rotation curves.

There is strong evidence that the masses of galaxies exceed the mass of the visible parts by some orders of magnitude. This goes back to Zwicky 1933 [107] who used the virial theorem to estimate the mass of galaxies in the Coma Berenices cluster and discovered that the mass exceeds luminosity mass by a factor of about 10^2. In current cosmological theory, this missing matter is identified with the invisible "dark matter" needed to make Keplerian motion fit the rotation curve. In the solution presented here, this extra matter is concentrated in the heavy rotating centre which controls the dynamics by inertial drag effects.

Assume that there is a standard background space (Minkowski or Schwarzschild space) and use an approximation to this background. Sciama's principle as discussed in Chapter 4 implies that the central rotating mass creates an inertial drag field dropping off like k/r, which causes inertial frames to rotate with respect to the background. With this assumption, it is not hard to solve the equations to find the tangential velocity in an equatorial orbit as a function of r (distance from the centre), and every equatorial orbit has the salient feature of observed rotation curves, namely a horizontal asymptote. This asymptote is *the same for all equatorial orbits* and hence any average over many orbits will also have this asymptote and this explains the observed rotation curve.

This provides strong evidence for the (weak) Sciama principle with inertial drag drop off asymptotically at k/r as promised at the end of Section 4.5.

5.1 The weak Sciama principle

Sciama's principle (Section 4.2) implies that the rotation of the local inertial frame (IF) is the sum

$$\sum_{Q} \frac{m_Q}{r_Q} \omega_Q$$

where the sum is taken over all (accessible) masses m_Q in the universe where m_Q is at distance r_Q and rotating with angular velocity ω_Q and the sum is suitably normalised .

For the purposes of this work, only the weak version is needed: (Section 4.5).

Weak Sciama Principle (WSP) *A mass M at distance r from P rotating with angular velocity ω contributes a rotation of $kM\omega/r$ to the inertial frame at P where k is constant.*

In the main application M will be the (heavy) centre of a galaxy, but the analysis applies to any axially-symmetric rotating body which does not need to be assumed to be heavy.

To fix notation, consider a central mass M at the origin in 3-space which is rotating in the right-hand sense about the z-axis (ie counter-clockwise when viewed from above) with angular velocity ω_0. Assume a flat background space-time, away from M, with sufficient fixed masses at large distances to establish a non-rotating IF near the origin, if the effect of M is ignored. Let P be a point in the equatorial plane (the (x, y)-plane) at distance r from the origin. The rotation of the inertial frame at P is given by adding the contribution from M to the contribution from the distant masses. Because P is near a large mass, it makes sense to normalise the sum as in Equation 4.3. This is equivalent to using a weighted sum, in other words the inertial frame at P is rotating coherently with the rotation of M by the average of ω_0 weighted kM/r and zero (for the distant fixed masses) weighted C say. Further normalise the weighting so that $C = 1$ (which is the same as replacing k/C by k) which leaves just one constant k to be determined by experiment or theory. The nett effect is a rotation of

$$(5.1) \qquad \frac{(kM/r) \times \omega_0 + 1 \times 0}{(kM/r) + 1} = \frac{A}{r + K} \quad \text{where } K = kM \text{ and } A = K\omega_0.$$

Note If the full Sciama principle is assumed and that $\sum_Q m_Q/r_Q = 1$ (Equation 4.2), which as was seen has some observational evidence to support it, then C and k are both 1 and $K = M$ and $A = M\omega_0$. However the choice of $k = 1$ is not relevant to the arguments presented in this or subsequent chapters. Nothing that is proved depends on knowing the exact relationship between K and M.

5.2   The dynamical effect of the inertial drag field

The key to the rotation curve is to understand the way in which the inertial drag field affects the dynamics of particles moving near the origin. For simplicity work in the equatorial plane. Assume that the IF at P (at distance r from the origin) is rotating with respect to the background with angular velocity $\omega(r)$ counter-clockwise. When computing rotation curves, the formula for $\omega(r)$ just found (5.1) will be used but for the present discussion it is just as easy to assume a general function. The IF at P can be identified with the background space, but it is important to remember that it is rotating. As remarked in Section 4.5 there is no sensible meaning to the centre of rotation for an inertial frame. Two rotations which have the same angular velocity but different centres differ by a uniform linear motion and inertial frames are only defined up to uniform linear motion. Thus it can be assumed for simplicity that all the rotations have centre at the origin. Then the IFs can be pictured as layered transparent sheets, each comprising the same point-set but with each one rotating with a different angular velocity about the origin. Each sheet corresponds to a particuar value of r. It is necessary to be very clear about the nature of motion in one of these frames. A particle moving with a frame (ie one stationary in that frame) has no *inertial velocity* and its velocity is called *rotational*. In general if a particle has velocity **v** (measured in the background space) then

$$\mathbf{v} = \mathbf{v}_{\text{rot}} + \mathbf{v}_{\text{inert}}$$

where its *rotational velocity* $\mathbf{v}_{\text{rot}}$ is the velocity due to rotation of the local inertial frame and $\mathbf{v}_{\text{inert}}$ is its *inertial velocity* which is the same as its velocity measured *in* the local inertial frame. Note that $\mathbf{v}_{\text{rot}} = r\omega(r)$ directed along the tangent.

The reader might find Figure 5.1 helpful at this point.

Inertial velocity correlates with the usual Newtonian concepts of centrifugal force and conservation of angular momentum.

As a particle moves in the equatorial plane it moves between the sheets so that a rotation about the origin which is rotational in one sheet becomes partly inertial in a nearby sheet. For definiteness, suppose that $\omega(r)$ is a decreasing function of r and consider a particle moving away from the origin and at the same time rotating counter-clockwise about the origin. The particle will appear to be being rotated by the sheet that it is in and this causes a tangential acceleration. This acceleration is called the *slingshot effect* because of the analogy with the familiar effect of releasing an object swinging on a string. But at the same time the particle is moving to a

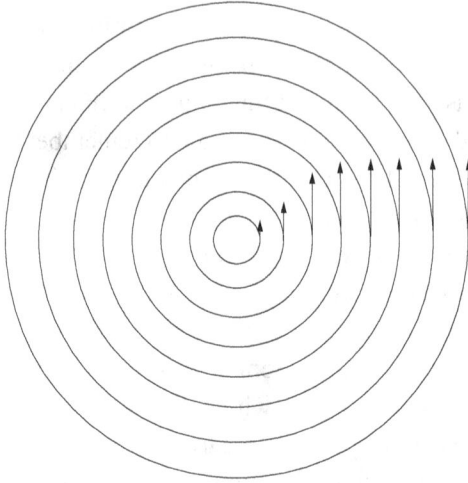

Figure 5.1: Rotational velocities in the inertial drag field near a rotating body

sheet where the rotation due to inertial drag is decreased and hence part of the tangential velocity becomes inertial and is affected by conservation of angular momentum which tends to decrease the angular velocity. These two effects balance each other out in the limit and this explains the flat asymptotic behaviour. Below this is proved analytically, but first, here is a metrical interpretation of the hypothesised inertial drag effect being used.

5.3   A metrical interpretation of inertial drag

Define an *inertial drag metric* by adding a variable rotation factor to a spherically-symmetric metric. The primary metrics of interest are obtained from the flat (Minkowski) metric and the Schwarzschild metric, but the proof of the rotation curve applies to any metric of this type. The inertial drag metric based on the Schwarzschild metric is likely to be close to the metric that will eventually be chosen if the conservative approach (cf Section 4.8) is generally adopted and serves to motivate the search for this metric.

Furthermore, as will be seen in the next chapter, a model for quasars based on the Schwarzschild metric successfully explains a good deal of the observations of these strange objects and this strongly suggests that this metric is a real reflection of reality at least in particular cases.

The most general spherically-symmetric metric can be written in the form:

(5.2) $$ds^2 = -B\,dt^2 + A\,dr^2 + r^2\,d\Omega^2$$

where A and B are positive functions of r and t on a suitable domain. Here t is time, r is "distance from the centre" (but see the note below) and $d\Omega^2$, the standard metric on the unit 2-sphere S^2, is an abbreviation for $d\theta^2 + \sin^2\theta\,d\phi^2$. Orient the 2-sphere so that the z-axis passes through it at the north pole where $\theta = \pi/2$. The (x,y)-plane (pasing through the origin and perpendicular to the z-axis) is the *equatorial plane* where (r,ϕ) are polar coordinates. The Schwarzschild–de Sitter metric is the case

$$B = \frac{1}{A} = 1 - \frac{\Lambda r^2}{3} - \frac{2M}{r}$$

with Λ and M constants. By Birkhoff's theorem (cf Section A.9) this is the only case where the metric satisfies Einstein's vacuum equations with cosmological constant in some region. In this case the metric is necessarily static in this region. The special cases $M = \Lambda = 0$ and $\Lambda = 0$ give the Minkowski and Schwarzschild metrics respectively.

Note It is important to observe that r is a coordinate *which is not precisely the same as distance in the metric*. It is chosen so that the sphere of symmetry at coordinate r has area $4\pi r^2$. Distance measured in the metric along a radius near this sphere is not the same as change in the coordinate r (this only happens if A takes the value 1 near the point under consideration).

The *inertial drag metric* is formed by adding a variable rotation about the z-axis. This is done by replacing ϕ by $\phi - \omega t$. The metric is no longer diagonal

(5.3) $$ds^2 = (-B + \rho^2\omega^2)\,dt^2 + A\,dr^2 + r^2\,d\Omega^2 - 2\rho^2\omega^2\,d\phi\,dt$$

where $\rho = r\sin\theta$.

If ω is constant this is the same metric viewed through rotating glasses, but the whole point is to allow ω to vary. Starting with the Schwarzschild–de Sitter metric and making this substitution with variable ω, gives a metric which no longer satisfies Einstein's vacuum equations: indeed the change made is the *metrical embodiment of the hypothesised inertial drag field*. It is not hard to see that the inertial frame at a point rotates about a line parallel to the z-axis with angular velocity the value of ω at that point. This is clear if ω is constant and in general, provided ω is continuous, it follows from the locality of inertial frames. So to fit with inertial drag as formulated in (5.1) it is necessary to set $\omega = A/(r + K)$ (at least in the (x, y)-plane). However it is easy to work with a general function ω and specialise when needed. The orbits of particles moving on geodesics in the equatorial plane will now be investigatied and, provided ω decreases like A/r as $r \to \infty$, the orbits will be found to fit observed rotation curves.

The reader might find Figure 5.2 helpful for visualising geodesics in the inertial drag metric and understanding the inertial drag effects. It shows geodesics on a typical cylinder $r = z = $const.

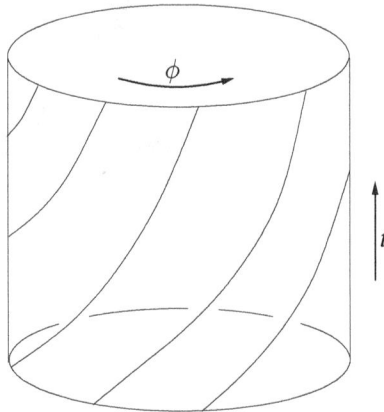

Figure 5.2: Geodesics on the cylinder $r = z = $const in the inertial drag metric

5.4 Conservation of angular momentum

Here now is the analytic derivation of the rotation curve (the relation between v and r) for the orbit of a particle in the equatorial plane moving with total velocity **v**, which has tangential component v (perpendicular to the line through the origin). Recall from the discussion in Section 5.2 above that there are two opposing effects at work: the slingshot effect, which tends to increase v with r and conservation of angular momentum which tends to decrease it. These two effects are calculated together. The proof works in any inertial drag metric, where the particle moves along a geodesic. (The special case of flat Minkowski space with inertial drag effects was motivated in Section 5.2.)

The derivation starts with a proof of conservation of angular momentum, which is a property of any system with a central force (or space-time geometry which simulates a central force). It is not restricted to Newtonian physics. The proof is adapted from Newton's proof of the equal area law for planetary orbits (which is exactly the same as conservation of angular momentum, see Figure 1.4). For the time being ignore ω (or set it equal to zero).

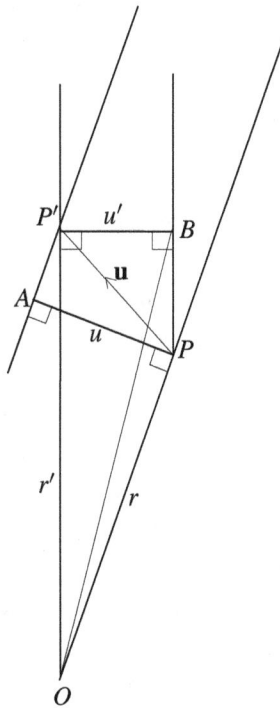

Figure 5.3: Proof of conservation of angular momentum

The idea is to replace the central force by a series of central impulses at equally spaced (small) intervals of time. Consider Figure 5.3. At a particular time the particle (of small unit mass) is at P and has just received a central impulse resulting in velocity $\mathbf{u}$. Its tangential velocity at P is $u = |AP|$. One small interval of time later the particle is at P' and receives another central impulse (along the line OP') which does not change its tangential velocity $u' = |P'B|$. But the triangle OPP' can be regarded as having base $r = |OP|$ and height u or base $r' = |OP'|$ and height u' hence

(5.4) $ur = u'r'$

in other words the angular momentum at P is the same as that at P'.

To obtain the result for an arbitrary continuous central force, take the limit of a sequence of central impulses. Note the proof does not use any property of the central force other than that it acts towards the centre. Nor does it assume that r represents a genuine distance in the metric under consideration. All that is needed is that Euclidean geometry correctly describes the relationship between r and distances perpendicular to radii near P and P' which is precisely how r was chosen.

5.5   The fundamental relation

Now reinstate w. Note that "force" in the model is a property of local space-time geometry. In the case that w is constant, the inertial frame (rotating with w) is the same as the unrotated case and in this frame the force is central. Therefore by locality it is central in the general case in the inertial frame. Therefore the proof just given makes sense in the inertial frame at P', in other words rotating with angular velocity $w' = w(P')$, though, as will be seen, in the limit the same result is obtained if it is assumed that the frame is rotating with angular velocity $w(P)$. To find the required relationship between v and r write v for the full tangential velocity at P and v' at P'. Since the frame is rotating at w', $v = u + w'r$ and $v' = u' + w'r'$. Write $v' = v + \delta v$, $u' = u + \delta u$, $r' = r + \delta r$ and $w' = w + \delta w$.

Since $ur = u'r'$ (Equation 5.4), substituting for u', v' and simplifying gives

(5.5) $u\,\delta r + r\,\delta u = 0\,.$

But

$$\delta u = u' - u = v' - \omega'r' - (v - \omega'r) = v' - v - \omega'(r' - r) = \delta v - \omega'\,\delta r$$

and substituting for $u, \delta u$ in (5.5) gives

$$(v - \omega'r)\,\delta r + r(\delta v - \omega'\,\delta r) = 0$$

which gives

$$r\,\delta v = 2r\omega'\,\delta r - v.$$

It is now possible to replace ω' by ω to first order (as forecast) and going to the limit yields the *fundamental relation* between v and r:

(5.6)
$$\boxed{\dfrac{dv}{dr} = 2\omega - \dfrac{v}{r}}$$

The fundamental relation can be understand intuitively as follows. The slingshot effect intuitively produces an acceleration $dv/dr = \omega$. On the other hand $v_{\text{inert}} = v - \omega r$ is the "inertial" tangential velocity (corrected for rotation of the local inertial frame) and therefore conservation of angular momentum produces a deceleration in v of v_{inert}/r or an acceleration $dv/dr = \omega - v/r$. Adding the two effects gives the relation.

5.6 Solving to find rotation curves

Given ω as a function of r, (5.6) can be solved to give v as a function of r. Rewrite it as

$$r\,\dfrac{dv}{dr} + v = 2\omega r.$$

The LHS is $d/dr\,(rv)$ and the general solution is

(5.7)
$$v = \dfrac{1}{r}\left(\int 2\omega r\,dr + \text{const}\right).$$

It is now clear that any prescribed differentiable rotation curve can be obtained by making a suitable choice of continuous ω.

Of interest here are solutions which, like observed rotation curves, are asymptotically constant and inspecting (5.7) this happens precisely when $\int 2\omega r\,dr$ is asymptotically equal to Qr for some Q and this happens precisely when 2ω is asymptotically equal to Q/r. This proves the following result.

Theorem   *The equatorial geodesics in the inertial drag metric (5.3) have tangential velocity asymptotically equal to constant Q if and only if ω is asymptotically equal to A/r where $Q = 2A$.*

5.7   The basic model

Now specialise to the case $\omega = A/(r + K)$ which gives the value of inertial drag formulated in (5.1).

From (5.7)

$$v = \frac{1}{r}\left(\int \frac{2Ar}{r+K}\,dr + C\right) = \frac{2A}{r}\left(\int 1 - \frac{K}{r+K}\,dr\right) + \frac{C}{r}$$

$$(5.8) \qquad = 2A - \frac{2AK}{r}\log\left(\frac{r}{K}+1\right) + \frac{C}{r}$$

where C is a constant depending on initial conditions. For a particle ejected from the centre with $v = r\omega_0$ for r small, $C = 0$, and for general initial conditions there is a contribution C/r to v which does not affect the behaviour for large r. For the solution with $C = 0$ there are two asymptotes. For r small, $v \approx r\omega_0$ and the curve is roughly a straight line through the origin. And for r large the curve approaches the horizontal line $v = 2A$. A rough graph is given in Figure 5.4 (left) where $K = A = 1$. The similarity with a typical rotation curve, Figure 5.4 (right), is obvious. Note that no attempt has been made here to use meaningful units on the left. See Figure 5.6 below for curves from the model using sensible units.

There are other shapes for rotation curves; see [96] for a survey. All agree on the characteristic horizontal straight line. Figure 5.5 is reproduced from [96] and gives a good selection of rotation curves superimposed. In Figure 5.6 is a selection of rotation curves again superimposed, sketched using Mathematica[1] and the model given here. The different curves correspond to choices of A, K and C. The similarity is again obvious. The units used differ. In the model given here natural units are used so that a velocity of .001 is 300km/s and a distance of 45,000 is 15Kpc approx.

---

[1]The notebook Rots.nb used to draw this figure can be collected from [4] and the values of the parameters used read off.

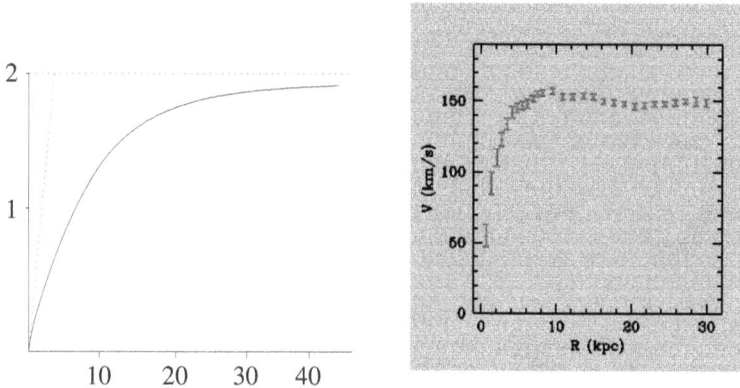

Figure 5.4: The rotation curve from the model (left) and for the galaxy NGC3198 (right) taken from Begeman [33]

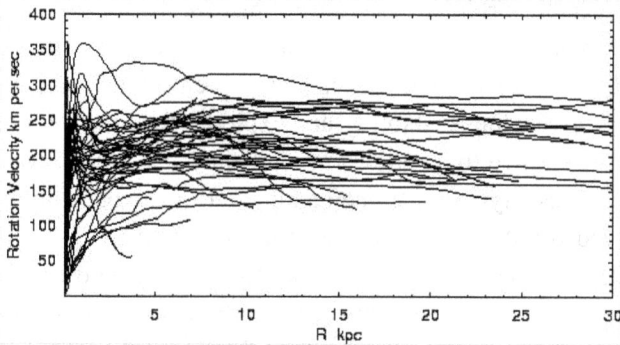

Figure 5.5: A collection of rotation curves from [96]

It is worth commenting that the observed rotation curve for a galaxy is not the same as the rotation curve for one particle, which is what has been modelled here. When observing a galaxy, many particles are observed at once and what is seen is a rotation curve made from several different rotation curves for particles, which may be close but not identical. So it is expected that the observed rotation curves have variations from the modelled rotation curve for one particle, which is exactly what is seen in Figures 5.4 (right) and 5.5.

The next chapter is devoted to the other main application of inertial drag, namely quasars. Then in Chapter 7 the analysis given here will be extended

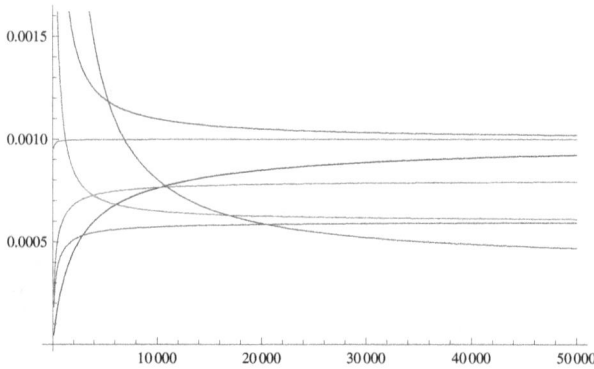

Figure 5.6: A selection of rotation curves from the model

to find equations for orbits in general (not just for the tangential velocity) and, using a hypothesised central generator, the spiral arm structure will be modelled as well. The basic idea is that the central mass accretes a belt of matter which develops instability and explodes feeding the roots of the arms. Stars are formed by condensation in the arms and move outwards as they develop. Thus a typical star is on a long outward orbit and the rotation curve observed for stars in a spiral arm is formed of many such similar orbits. But this full picture is not necessary to explain the observed rotation curves, since the tangential velocity for all orbits has the same horizontal asymptote. Chapter 7 is more specific about the size of the central mass in a galaxy. These vary from 10^9 to 10^{14} solar masses with the range 10^9 to 10^{11} corresponding to so-called "active galaxies" and the range 10^{11} to 10^{14} to full-size spiral galaxies. The central masses for the curves in Figure 5.6 vary from 3×10^{11} to 10^{14} solar masses and it is useful to know that a mass of 1 in natural units is 3×10^{11} solar masses.

5.8 Postscript

As remarked earlier, the effect described in this chapter is independent of mass. However for rotating bodies of small mass the effect is unobservably small. For example the sun has $K \approx 3$km, assuming $K = M$, and $\omega_{Sun} = 2\pi/25$ days. Thus the asymptotic tangential velocity $2A = 2K\omega_{Sun}$ is 6km per 4 days or .06 km per hour.

Chapter 6

Quasars

quasar
kweɪzaː, kweɪsaː,
noun Astronomy
noun: quasar; plural noun: quasars

> a massive and extremely remote celestial object, emitting exception-
> ally large amounts of energy, which typically has a starlike image
> in a telescope. It has been suggested that quasars contain massive
> black holes and may represent a stage in the evolution of some
> galaxies.

Origin
1960s: contraction of quasi-stellar.

Google dictionary definition (October 2017)

Quasars were first observed in the 1960's. Through a telescope they appear
to be stars but they exhibit strange features not shared by ordinary stars.
They have spectra which often appear to be hugely redshifted and they
vary irregularly with time scales that range from hours to months. Early in
the study of quasars a heated controversy raged about these huge redshifts.
Are they cosmological due to the expansion of the universe? or are they
gravitational due to the near presence of a massive object (eg a black hole)?

The cosmological explanation implies that quasars with large redshifts are extremely distant objects with truly phenomenal power outputs which are very hard to explain. By constrast the gravitational explanation allows the possibility that they are modest size objects, not too distant and with easily modelled power outputs. As can be seen from the Google definition, the cosmological explanation is the currently accepted one. This is despite some incontrovertible evidence in the form of observations of Halton Arp and others [31, 45] that quasars are often closely associated with galaxies with the redshift for the quasars significantly higher than that for the associated galaxies, a striking example of which was previewed in Section 3.4 and is reproduced in again Figure 6.1.

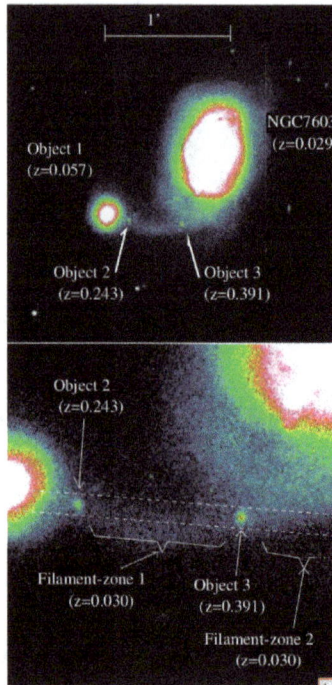

Figure 6.1: NGC 7603 and the surrounding field. R-filter, taken on the 2.5 m Nordic Optical Telescope (La Palma, Spain). Reproduction of Figure 1 of [68]

This example contains two Arp quasars (objects 2 and 3) strongly associated with a galaxy (and companion) both of lower redshift. Lopez Corredoira and

Gutierrez [68] report $z = 0.0295$ and $B = 14.04$ mag for the main galaxy, NGC 7603 and comment: "A fact that attracted attention is its proximity to NGC 7603B (Object 1 hereafter), a spiral galaxy with higher redshift $z = 0.0569$, moreover a filament can be observed connecting both galaxies. They also found two objects superimposed on the filament with redshifts 0.394 ± 0.002 and 0.245 ± 0.002 for the objects closest to and farthest from NGC 7603, Objects 3 and 2, respectively. B-magnitudes corrected for extinction (due to the filament) are respectively 21.1 ± 1.1 and 22.1 ± 1.1."

It is commonsense that the alignments seen in Figure 6.1 are not due to chance and there are many similar such in Arp and other's observations [31, 45]. Objects 2 and 3 have cosmological redshift around $z = 0.030$ (for the filament) and the remainder must be instrinsic (presumably gravitational). [As remarked in Section 3.4, as often happens when a consensus view is challenged by direct evidence, the evidence is ignored and the challenger discredited — Arp was sidelined by the mainstream cosmological community and denied observation time on the big telescopes. The author's hope is that this book will rehabilitate Arp's reputation (unfortunately posthumous).]

The purpose of this chapter is to explain how the same inertial drag phenomenon used in the last chapter to model rotation curves can be used to restore a sensible explanation for these observations and to establish a simple model for quasars with modest power output that explains all the observations.

6.1 Angular momentum and inertial drag

As explained in the introduction, the strongest argument supporting the current consensus view (that redshift in quasars is cosmological) comes from a consideration of angular momentum. Assume that a quasar contains a large central mass (presumed to be a black hole) and that its energy production is due to accretion from the surrounding medium. Particles fall into the gravitational well of the central mass and the gravitational energy is released by interaction between different infalling particles. Now given a small but very heavy object, a particle approaching with a small tangential velocity will have its tangential velocity magnified by conservation of angular momentum and there will be a radius of closest approach. It is very unlikely to actually fall into the central gravitational well, and the same thing happens for the full flow of infalling matter from the surrounding medium, which will typically have a nonzero angular momentum around the black hole. (See the Michel quote in Section 3.5.) This gives an obstruction to accretion, which was found not long after quasars were discovered, and led to the subject being dominated by the theory of accretion discs.

But now assume that the central mass is rotating and that the infalling particle is in the equatorial plane. Take inertial drag effects into consideration. From (5.7) the angular momentum per unit mass of the particle can be read as $vr = \int 2\omega r \, dr + \text{const}$. This is the *apparent* angular momentum (as calculated by a distant viewer). To find the *true* angular momentum (ie as measured in the local inertial frame) replace $\mathbf{v}$ by $\mathbf{v}_{\text{inert}}$, which entails subtracting ωr from v, so the true angular momentum (per unit mass) is

(6.1)
$$vr - \omega r^2 = \int 2\omega r \, dr - \omega r^2 + \text{const}$$

By suitable choice of the integration constant, there are solutions with low angular momentum (either true or apparent) for r small and significant angular momentum for larger r and it follows that the effect of the inertial drag is that *the rotating body can absorb angular momentum.* And notice that this holds for almost any dynamical theory, in particular general relativity. (It also works with almost any nontrivial function ω.)

Now if angular momentum can be nullified by central rotation, then it does not force the existence of an accretion disc and a simple spherically-symmetric model for accretion can be used. Here is another description of the effect being used here which gives further information. The formula for ω (Equation 5.1) implies that IFs near the origin rotate at roughly the same rate, in other words all fit with a plate-like rotation. If the speed of this rotation is the same as the effective rotation of the infalling matter then the latter rotation will be "rotational" (due to the rotation of the IF) and not "inertial". Thus conservation of angular momentum (which acts only on inertial velocity) will not change it and the inflow will be radial in the local inertial frame. Moreover there is a feedback effect working in favour of this. If the incoming matter has excess angular momentum, then it will tend to contribute to the central rotation which therefore changes to increase the inertial drag effect until the two balance again. Conversely, if there is a shortfall, the black hole will slow down. In other words, once locked on the ambient conditions that allow the black hole to accrete, there is a mechanism for maintaining that state.

At this point recall that given a black hole with angular velocity ω and angular momentum Ω then $\Omega = M\omega r_{\text{eff}}^2$ (4.6) where r_{eff} is the effective radius of the black hole, assumed to be small but not zero. Thus a very small change in angular momentum corresponds to a reasonable change in the angular velocity. This makes the locking effect described above more responsive and effective.

If the incoming matter has angular momentum about a different axis than the rotation axis for the black hole, then a similar feedback effect will cause the black hole rotation axis to change into alignment with axis for the incoming matter.

The conclusion is that the angular momentum obstruction for accretion can effectively be ignored and a spherically symmetric accretion model used. A suitable model based on the Schwarzschild metric is studied in joint work with Robert MacKay and Rosemberg Toala Enriques [88]. In this model gravitational redshift can take arbitrary values

(6.2) $$z = 1.27 \times 10^7 \, \mathcal{M}^{-1} \, n^{-1} \, T^{1.5} \, [1/(2X)]$$

where $\mathcal{M}$ is the black hole mass in solar masses, n is density of the ambient gas/plasma in number of particles per cubic metre (assumed to be Hydrogen atoms or protons) and T is temperature in degrees Kelvin. X is an absorption factor which can be taken to be $1/2$ (ie ignore the factor in square brackets). The full technical details of the model (hereafter called the "three-author model") are given in Appendix C, where Equation (6.2) is proved. Notice one important point about this equation. The mass of the black hole $\mathcal{M}$ appear inverted so that (other parameters being equal) *redshift decreases with black hole mass*. This was observed directly by Arp and caused him to invent some fantasy physics to explain it because he was not aware of the considerations discussed here.

Also in the appendix are many worked examples including NGC7603 and associated objects. One particular quasar is worth mentioning here because of the (in fact false, as will be seen) importance that it has for the Milky Way, namely Sagittarius A*. This quasar is regarded as problematic by the mainstream quasar community because its level of radiation is 8 orders of magnitude below the Eddington limit. It is suggested here that this is due to a very high redshift, $z = 10^4$, which causes the power to be attenuated by the square of this, namely 10^8. Full details and supporting evidence from the luminosity graph can be found at the end of Section C.6.

This chapter finishes with an outline of this three-author model and discussion of evidence and previous work on quasars. Note that this chapter and the related appendix use MKS units and not the natural units used in other chapters.

6.2   Outline of the three-author model

As remarked earlier this model for black hole radiation (aka quasar radiation) is spherically-symmetric, fully relativistic and based on the Schwarzschild metric. There are fully relativistic Schwarzschild black hole models to be found in the literature, for example the models of Flammang, Thorne and Zytkow [43] quoted by Meier [73, page 490]. But the significance of these models, and in particular their redshift, has been ignored, presumably because of the angular momentum obstruction discussed above. Thus the excellent fit with observations that is found has been overlooked.

In the three-author model, black holes radiate by converting the gravitational energy of incoming matter into radiation and, since only a fraction of the available energy is radiated back out, they accrete mass and grow over time. There will be a good deal more to say about this growth in later parts of the book. It is highly suggestive of a life-form.

The basic set-up considered is a black hole floating in a gas of Hydrogen atoms (the *medium*), which might be partially ionised (ie form a plasma), with the radiation coming from accretion energy. Matter falls into the black hole and is accelerated. Interaction of particles near the black hole changes the "kinetic energy" (KE) of the incoming particles into thermal energy of the medium and increases the degree of ionisation. The thermal energy is partially radiant and causes the perceived black hole radiation.

Kinetic energy is not a relativistic concept as it depends on a particular choice of inertial frame in which to measure it. It is for this reason that it has been placed in inverted commas. Nevertheless, it is a very useful intuitive concept for understanding the process being described here.

The following simple considerations suggest that most of the KE of the infalling matter is converted into heat and available to be radiated outwards. A typical particle is very unlikely to have purely radial velocity. A small tangential velocity corresponds to a specific angular momentum. As the particle approaches the black hole, conservation of angular momentum causes the tangential velocity to increase. Thus the KE increase due to gravitational acceleration goes largely into energy of tangential motion. Different particles are likely to have different directions of tangential motion and the resulting melée of particles all moving on roughly tangential orbits with varying directions is the main vehicle for interchange of KE into heat and hence radiation. Very little energy remains in the radial motion, to be absorbed by the black hole as particles finally fall into it. Thus the overall radial motion of particles is slow. In terms of the models of [43], the "breeze solutions" for radial flow [73, Figure 12.2, page 489] are being used. Far away from the black hole, where density is close to ambient density, and therefore low, this process converts angular momentum into radial motion with little loss of energy and serves to allow the plasma to settle into the inner regions, where the density is higher and the particle interactions generate heat and radiation.

6.3   Three important spheres

For simplicity of exposition now assume that the medium is a Hydrogen plasma and the heavy particles are therefore protons. This is true in the higher temperature parts of the model, for example once the Eddington sphere is reached, see below. But there is no material difference if the medium is in fact a partially ionised Hydrogen gas.

Observations of quasars often show the presence of other atomic material in the radiation zone so that this simplifying assumption may need revision at a later stage.

There are three important spheres. The outermost sphere is the *Bondi sphere* of radius $B = 2GMm_H/3kT$ defined by equating the average velocity of protons in the medium with the escape velocity at radius B. Here M is the black hole mass, G is the gravitational constant, k is Boltzmann's constant, T is temperature and m_H is the mass of a proton.

The significance of the Bondi sphere is that protons in the medium are trapped (on average) inside this sphere because they have KE too small to escape the gravitational field of the black hole. The mass of matter per unit time trapped in this way is called the *accretion rate A* and can be calculated as

$$(6.3) \qquad\qquad A = 2B^2 n \sqrt{2\pi kTm_H}$$

where n is the density of the medium (number of protons per unit volume).

Details for these calculations are given in Section C.1.

Proceeding inwards, the next important sphere is the *Eddington sphere* of radius R which is defined by equating outward radiation pressure on the protons in the medium with inward gravitational attraction from the black hole. More precisely, the outward radiation pressure acts on the electrons in the medium which in turn pull the protons by electrical forces. This is the same consideration as used to define the Eddington limit for stars and this is why the same name has been used. At the Eddington sphere the gravitational pull on an incoming proton is balanced by the outwards radiation pressure (mediated by electrons) and, assuming the radiation pressure is just a little bigger, the acceleration of the incoming proton is replaced by deceleration and the KE of infall is absorbed by the medium and available to feed the radiation. It is a definite hypothesis that there is an Eddington sphere, but the final model that is constructed using this hypothesis does fit facts pretty well, and this justifies it.

It is helpful to think of the Eddington sphere as a transition barrier akin to the photosphere of a star. Indeed the Eddington radius R is also the radius at which photons get trapped in the medium and for this reason is also known as the trapping radius. This can be seen by thinking of the forces that define it the other way round. The incoming matter flow exerts a force on the outward radiation and when these two are in balance, the outward radiation is stopped and photons are trapped.

Thus at the Eddington sphere two things are happening: the infalling protons are stopped and their KE released into the general pool of thermal energy and the outward flow of radiation is also stopped. Thus radiation from the black hole is generated by activity in the close neighbourhood of the Eddington sphere and this is the place where redshift of the outward radiation due to the gravitational pull of the black hole arises.

The region outside the Eddington sphere is optically thin whilst the region inside is optically thick. The radiation that is emitted comes from a narrow band near the Eddington sphere and which is all at roughly the same distance from the central black hole. This allows the radiation to exhibit a consistent redshift.

Precise formulae that determine the Eddington radius in terms of the other parameters are given in Section C.2.

The final sphere is the familiar Schwarzschild sphere or event horizon of radius $S = 2GM/c^2$ where M is the black hole mass.

The region between the Schwarzschild and Eddington spheres is called the *active region* and the region between the Bondi sphere and the Eddington sphere, the *outer region*. A simplifying assumption is made that nearly all the KE that powers the black hole is released in the active region. This means that any KE turned into heat by particle interaction in the outer region is ignored. This is justified by the fact that this region has low density, close to the ambient density, so that most particle interactions are between particles sufficiently far apart to conserve kinetic energy. It is useful to think of this region as a "settling region" where angular momentum is converted into radial motion, allowing the plasma to settle towards the active region. See also the discussion below Equation (C.6) and in Section C.7.

One other simplifying assumption is made: it is assumed that there is no significant increase in temperature near the Bondi sphere due to the black hole radiation, ie T is the ambient temperature.

6.4   Previous work on quasars and gravitational redshift

This chapter finishes with a review of the historical reasons for abandoning the idea that quasars might have significant instrinsic (gravitational) redshift and why they do not apply to the model.

The principal reason (angular momentum) has already been fully explained. There are four main further reasons:

(1)   Redshift gradient (see the discussion in [80] on pages 3–4)

If redshift is due to a local mass affecting the region where radiation is generated, then the gravitational gradient from approach to the mass would spread out the redshift and result in very wide emission lines. This effect is called "redshift gradient".

In the model, although the energy production takes place throughout the active region, the emitted radiation is generated only at (or near) the Eddington sphere which is all at the same distance from the central mass and subject to the same redshift. Thus the model has the observed property that emission lines are moderately narrow.

(2)   Forbidden lines (cf Greenstein–Schmidt [47])

Many examples of black hole radiation show so-called forbidden lines, which can only be produced by gas or plasma at a fairly low density. The assumption that *all* the radiation is produced by a low density region leads to an implausibly large and heavy mass (see [47, page 1, para 2]).

In the three-author model, the region directly adjacent to the Eddington sphere is at roughly ambient density which is, in all examples that are examined in the appendix, low enough to support forbidden lines (more details on this will be given in Section C.5). A narrow shell of low density near the Eddington sphere is excited by the radiation produced at the sphere and produces radiation in turn. It is here that forbidden transitions take place and result in the observed forbidden lines.

(3) Mass and variability problems (cf Greenstein–Schmidt [47], Hoyle–Fowler [56])

The mass problem is a rider on the forbidden line problem but also applies to attempts at models for gravitational redshift without significant redshift gradient. As remarked above, assuming that all the radiation is produced by a low density region leads to an implausibly large and heavy mass. The same thing happens if one tries to produce a region with sufficient local gravitational field to provide a base for the radiation production, without redshift gradient, as for example in Hoyle and Fowler [56]. This problem is compounded by the fact that quasars typically vary with time scales from days to years. For variability over a short timescale, a small production region is needed (significantly smaller than the distance that light travels in one period).

It is worth remarking in passing that this problem is unresolved by the current assumption that all quasar redshift is cosmological. This implies that quasars are huge and very distant so that special (and unnatural) mechanisms are invoked to explain variability.

In the three-author model, the size of the radiation producing region is small enough. The black hole sizes that fit observations are in the range 10^3 to 10^8 solar masses. For quasars with significant intrinsic redshift, the radius of the Eddington sphere has the same order of magnitude as the Schwarzschild radius, and for 10^8 solar masses this is 3×10^{11} metres or 10^3 light seconds or about 20 light minutes. Thus the natural mechanism for variability, namely orbiting clouds or more solid bodies causing periodic changes in observed luminosity, fits the facts perfectly.

It is also worth observing here that there is a quite remarkable paper of M R S Hawkins [52], which proves an apparently paradoxical result, namely that a certain sample of quasars exhibits redshift without time dilation. The paradox arises from the fact that redshift and time dilation are identical in general relativity. Indeed they are identical in any theory based on space-time geometry. What Hawkins actually finds is a sample of quasars with varying redshift for which the macroscopic variation in light intensity does not correlate with the redshift. The resolution of the paradox is that the mechanism that produces the redshift and the mechanism which causes the variability are not subject to the same gravitational field. This is precisely how the model works. The redshift is caused by the central black hole and the variability is caused by orbiting clouds etc, much further out, and in a region of lower redshift. For more detail on the Hawkins paper and its meaning see Section 9.8. Properly understood, the paper proves conclusively that quasars typically have intrinsic redshift.

(4)   Statistical surveys

Stockton [95] is widely cited as a proof that quasar redshift is cosmological. He takes a carefully selected sample of quasars and searches for nearby galaxies within a small angular distance and at close redshift. Out of a chosen sample of 27 quasars, he finds a total of 8 which have nearby galaxies with close redshifts. He assumes that all these quasars have significant intrinsic redshifts and are therefore not actually near their associated galaxies. He then calculates the probability of one of these coincidences occurring by chance at about $1/30$, and concludes that the probability of this number of coincidences all occurring by chance is about 1.5 in a million.

The conclusion he draws is that all quasar redshift is cosmological.

The fallacy is obvious from this summary. It may well be that many of the quasars in the survey do not have significant intrinsic redshift and therefore some of these coincidences are not chance events. As can be seen from Equation (6.2) the three-author model allows the gravitational redshift of a quasar to vary from near zero to as large as you please. Roughly speaking, redshift is small (orders of magnitude smaller than 1) if the mass is big or the medium is dense and cold. Conversely, with a small mass and a hot thin medium, the redshift can be several orders of magnitude greater than 1. There is a natural progression for a quasar, as it accretes mass and grows heavier, to start with a very high gravitational redshift and gradually evolve towards a very low one. Without a sensible population model for quasars, it is difficult to comment on the number of coincidences that Stockton finds, but it is highly plausible that heavy quasars (with low gravitational redshift and central masses of say 10^7 to 10^9 solar masses) gravitate towards galactic clusters and therefore have nearby galaxies at a similar cosmological redshift. This would provide a natural framework for the Stockton survey within the model.

Stockton does discuss the possibility that quasars may have both small and large intrinsic redshifts (see [95, page 753, right]), but the discussion is marred by assuming that the two classes must be unrelated objects. The three-author model has a natural progression between the two classes.

There is a more modern survey by Tang and Zhang [98] which also claims to prove that all quasar redshift is cosmological. But examining the paper carefully, what is actually proved is that some particular models for quasar birth and subsequent movement are incompatible with observations. To comment properly on this paper a good population model for quasars would again be needed. But it is worth briefly mentioning that at least one of their models (ejection at 8×10^7 m/s from active galaxies with a lifespan of 10^8 years) does fit facts fairly well, see [98, Figure 1, page 5]. The ejection velocity is implausibly large, but the lifespan could easily be 50 times larger allowing for a plausible ejection velocity of say 10^7 m/s and a better fit with the data.

Finally, there is another interesting argument given by Wright [105] "proving" that quasar redshift is all cosmological from details of the spectra. This is the Lyman-alpha-forest argument. The observations he cites give useful information about the outer region. This, and the fallacy in the argument, will be discussed near the end of Appendix C in Section C.7.

The story continues in Appendix C where full technical details of the three-author model and the fit with data can be found.

But to finish this chapter here are some comments on quasar growth. It has been seen that quasars grow by accretion and lose their intrinsic redshift (as observed by Arp, but explained using non-standard physics). If, as Arp suggests, they are ejected from mature galaxies, then there is a natural way to think of them as young galaxies. As they grow and gain mass, they will take on more and more features of active galaxies and finally develop into mature spiral galaxies (discussed in the next chapter). As a highly speculative example, the grouping of four objects (two galaxies and two quasars) seen in Figure 6.1 could be a "family" group: two adults and two children. Indeed the quasar–galaxy spectrum has all the appearances of forming the dominant lifeform for the universe. This topic is taken up again in Sections 7.2 and 9.8.

Chapter 7

Spiral structure

The chapter combines ideas from the last two chapters to give a complete description of the dynamics of spiral galaxies.

7.1 Introduction

Spiral galaxies (Figures 7.1, 7.3, 7.7) are surely the most beautiful objects in the universe and it comes as a shock to find that there is no proper theory for their structure in current cosmology.

The main problem arises from the assumption that stars move on roughly circular orbits. In order for a spiral structure to be maintained stably over several revolutions, with all stars moving on circular orbits, it is necessary for tangential velocity to be roughly proportional to distance from the centre which means that the rotation curve is far from the one observed, Figure 7.2 (left). This problem is known as the "winding dilemma". In order to solve this problem conventional cosmology proposes that the spiral arms are not real but virtual. It proposes that they are in fact "standing waves" or "density waves", Figure 7.2 (right). Although this theory gives plausible spirals, the nature of the arms in real galaxies, Figure 7.3 (right) or Figure 7.7 (left) does not fit it at all. Real arms are composed of a spiral curve of intense star producing regions and associated high luminosity short-life stars, see for

Figure 7.1: M83 Southern Pinwheel. Left: image from European Southern Observatory [6]. Right: close-up from Hubble site [7]

example the close up of M83 from the Hubble site Figure 7.1 right. There is no trace of the orbits that are supposed to form the density wave outside the actual spiral arm. To be a little fairer to the standing wave theory, there is a rider to the theory which suggests that a shock-wave effect causes short-life stars to appear as the standing wave moves. Indeed it is clear from any galactic picture that the main luminosity of typical spiral arms comes from high luminosity short-life stars, but the short-life stars produced by a shock wave would last long enough to blur the arms and the pictures are quite clear: no such blurring occurs. Moreover, this model begs the question of where the continuous supply of pre-stellar material comes from to support this creation process.

In this chapter a quite different solution to this problem is proposed. The idea is, instead of assuming that stars move on roughly circular orbits, to assume that they move outwards along the arms as they rotate around the centre. Thus the familiar spiral structure is like the visible spiral structure in a Catherine wheel, the arms being maintained by stars moving along them, and there is no need for any special pleading to explain the observed structure. Moreover, the motion can be modelled and the stable spiral structure demonstrated . The model has one crucial feature in common with the standing wave theory in its shock-wave version: the visible spiral structure does consist mainly of short-life stars and star-producing regions

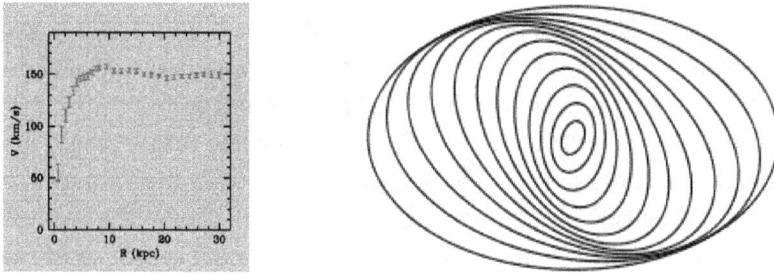

Figure 7.2: Left: the rotation curve for the galaxy NGC3198 reproduced from [5] (sourced from Begeman [33]). Right: the standing wave theory, reproduced from wikipedia

and this common feature is crucial for accurate modelling because it allows for elapsed time along the arms of about 10^8 years which fits the models constructed here. More details on this are given in Section 7.6. The problem of supply of pre-stellar material is solved in the model by continuous replenishment from the centre of the galaxy.

Two assumptions are needed, the first of which was anticipated in Chapter 5 to explain the observed rotation curve, namely that the centre of a normal spiral galaxy such as the Milky Way contains a hypermassive black hole, of mass 10^{11} solar masses or more. The second assumption is that this black hole is ringed by an accretion torus of a very precise type, which is called the *generator* or *belt* and which is responsible for generating the streams of material which feed and maintain the spiral arms. The belt is an example of the accretion structures hypothesised for the nuclei of "active" galaxies (for which central super-massive black holes, of 10^8 to 10^{10} solar masses, have been directly observed) and used to explain their observed radiation. "Active" has been placed in quotation marks because of one of the main theses of this book, that all galaxies are active: the activity of a spiral galaxy is responsible for its spiral structure; indeed there is no distinction between active galaxies and "normal" spiral galaxies. The real difference is that the central black hole in a normal galaxy is masked from view by the matter which is trapped in accretion structures near it, the most prominent of which is the central bulge. The spectrum of black hole based objects will be discussed further in Chapter 8.

The assumption of a hypermassive central black hole in a spiral galaxy directly contradicts current beliefs of the nature of Sagittarius A* and this problem together with other observational matters will be dealt with in the next chapter (Section 8.3). Very briefly, SgrA* and the stars in close orbit around it form an old globular cluster near the end of its life with most of the matter condensed into the central black hole. It is not at the centre of the galaxy but merely roughly on line to the centre and it is about half-way from the sun to the real galactic centre which is invisible to us.

As remarked above, in the new model for galactic dynamics proposed here, young stars in a galaxy are moving outwards as well as around the centre. This general outward movement has not been observed, although there are some old observations of Oort, Kerr and Westerhout [79] which show outward movement in gas clouds but which are generally misinterpreted (see Section 8.1). Indeed, early observations of Lindblad, using Shapley's maps of globular cluster, suggested that stars in the neighbourhood of the sun move on circular orbits (see [35, page 16]) and this has created an idée fixe that all stars in galaxies move on roughly circular orbits with any contrary observations explained away on an ad hoc basis. In the model proposed here, motion of stars is far from Keplerian, being strongly controlled by inertial drag effects from the (rotating) centre. The result is that the outward progress takes a very long time — commensurate with the lifetime of a star — and hence the outward velocity, far out from the centre where the Sun lies, is rather smaller than (about one tenth of) the observed rotational velocity. Thus the new model is consistent with the Lindblad observations. For more detail here, see the analytic models constructed in Section 7.6.

The general picture which emerges is of a structure stable over an extremely long timescale (at least 10^{12} years) with stars born and aging on their outward journey from the centre and returning to the centre to be recycled with new matter to form new solar systems. The tentative suggestion is that galaxies have a natural lifetime of perhaps 10^{16} years with the universe considerably older than this. The consequences of these suggestions for cosmology as a whole will be discussed in the next chapter; here note that the theory of galactic dynamics presented in this chapter does not depend on this timescale. Indeed it could at a pinch be consistent with the current standard model for the universe as a whole starting with the Big Bang. But the author's opinion is that the Big Bang theory is a serious mistake. For more detail here see Section 9.1 .

Figure 7.3: M101 (left) and NGC1300 (right): images from the Hubble site [7]

7.2 The generator

The full dynamics of spiral galaxies will be developed in the next and following sections, but first here is an outline of the proposed generator for the spiral arms. The story is part of the main story of the book, namely the quasar–galaxy spectrum, which will be taken up again in Section 9.8 and following sections. The keys to understanding the generator are the familiar ones: angular momentum and inertial drag. Previous chapters have covered the start of the story: inertial drag effects allow a black hole (aka quasar) to cancel out the angular momentum obstruction to accretion and feed on the surrounding medium and hence grow in size. As it grows the mass increases and its intrinsic (aka gravitational) redshift decreases. For a very small quasar such as SgrA* of mass $10^{6.6}$ solar masses, the intrinsic redshift can be very large (in this case a figure of $z = 10^4$ is indicated by observations), but for a larger quasar (of size say 10^8 solar masses) $z = 0.05$ is more typical, but there is a huge variation, see the tables at the end of Section C.6.

As the mass grows and the instrinsic redshift decreases, the simple spherical accretion model, described in Chapter 6 and Appendix C, breaks down because the accretion rate is too great for smooth accretion to take place. The outer settling region develops instabilities; there is evidence for this starting in the so called "Lyman-alpha forest" (see the end of Section C.6). The flow accumulates near the Eddington sphere choking the inflow. A rotating toroidal accretion structure (the belt), similar to the conventional theory, forms Figure 7.4 (left).

Notice that as the material in orbit around the black hole grows in mass and extent, it increasingly masks the central black hole and the virial theorem typically used to estimate this mass becomes less useful. This point in the spectrum is where "active" galaxies (of mass in the range $10^8 - 10^{10}$ solar masses) start.

The quasar starts to produce explosive outflow (jets) and to morph into an active galaxy. The angular momentum locking effect, described in Section 6.1, is now no longer stable because the jets carry away angular momentum and cause the whole system to rotate in the *opposite direction* to the rotation of the belt. So the central black hole is rotating to the left (say) and a surrounding belt is rotating to the right Figure 7.4 (right). (Note that throughout this chapter anticlockwise or positive (positive value of $\dot{\theta}$) is used in all illustrations for the main rotation of the central body; the belt has negative (clockwise) rotation.) This implies that the angular momentum in the inertial frames is augmented by the inertial drag effects described earlier and the effective energy in the belt similarly augmented. Thus energy is being fed into the belt structure directly from the black hole itself. There are also two other sources of energy for the belt: accretion energy as for any quasar and energy (heat) caused by interference due to inertial drag between layers: a sort of "friction" effect. With all this energy going into the belt, it becomes extremely hot and a plasma of quarks forms nearest the centre, condensing into a normal plasma of ionised H and He nuclei, with a trace of Li, further out. Conditions here are similar to those hypothesised to have occurred just after the Big Bang and the resulting mix of elements is the same. The energy results in explosions causing the jets mentioned already. As mass increases the jets become massive and permanently established and manifest themselves as the familiar spiral arms of the galaxy as explained below.

At the same time there is a build-up of matter trapped near the central black hole, visible as the familiar bulge, which totally masks the black hole, and

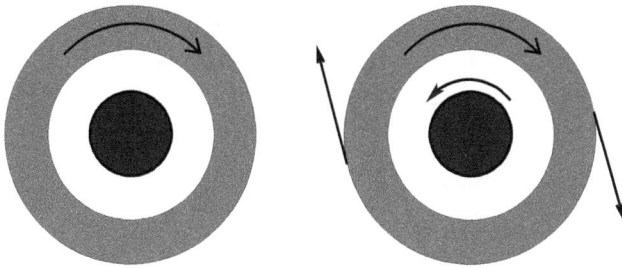

Figure 7.4: Left: the belt rotates clockwise. Right: ejected matter causes the whole system to rotate anti-clockwise.

the fiction that the central black hole of the Milky Way could be only $10^{6.6}$ solar masses is not obviously wrong (though it is completely incompatible with the dynamics presented in this chapter).

Once the central black hole starts rotating, the inertial drag effects calculated in Sections 5.2 and 7.4 come into play and, further out, matter ejected from the centre starts to rotate *with* the hole and against the rotation of the belt. Matter is lost from the outer regions and, if ejected from the centre fairly slowly so that the inertial drag effect dominates, carries away angular momentum of the opposite sign, Figure 7.5. There is a stable situation in

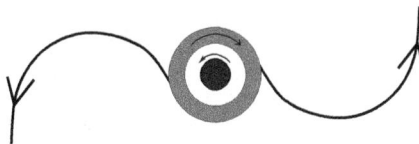

Figure 7.5: Inertial drag carries ejected matter anticlockwise and a balance is reached. Arms form.

which the loss of angular momentum in both directions is in balance: highly energetic particles ejected from the belt are not strongly affected by inertial drag effects and carry away clockwise angular momentum; less energetic particles are affected and carry away anticlockwise angular momentum. This balancing effect is why there is strong stability in the limiting tangential velocity, in other words why rotation velocity is roughly constant over all galaxies observed, Figure 5.5. Note that Figure 7.5 shows orbits *not* arms.

It should be compared with Figure 7.9 left. Inertial drag causes the roots of the arms to appear to precess clockwise and the snapshot of orbits that is seen has the familiar spiral form (as demonstrated in the Mathematica drawn figures in Section 7.6).

Energy is lost from the black hole because of the matter ejected from the belt, but energy is recovered by matter falling into the active region near the black hole so that the whole structure is stable over an immense timescale. In the next chapter compatibility of this model with the timescale of the Big Bang is discussed. This global loss of energy implies that there is no further growth in general, indeed there is now steady loss due to radiation energy and matter lost to the system, to balance accretion.

The spiral arms form as follows. The explosions from the belt mentioned above are the mechanism which feeds the spiral arms. These do not occur in random places: most normal galaxies have a pronounced bilateral symmetry with two main opposing arms (eg Figures 7.1, 7.3 and 7.7). There is no intrinsic reason for this to happen, but it is a stable situation. Once two arms have formed, then the gravitational pull of these arms will form bulges at the roots of the arms and encourage explosions there to feed the arms. The bilateral symmetry arises because the bulges are tidal bulges, caused by the pull of the nearby spriral arms, and tidal bulges always have bilateral symmetry. This tendency to bilateral structure is weak and looking at a gallery of galaxies many examples where it fails to form or where other weak arms have formed as well as the two main arms can be found.

Notice that ejection from the belt is generally in the direction of the belt rotation, which is opposite to the direction of the black hole rotation and this as will be explained later is why the roots of the arms generally have a noticeable offset (see the galaxy examples referred to above). To be precise, there is a constant C in the model which sets the tangential velocity at the root of the arms, and with this set negative, the arms are offset, cf Section 7.6.

How the structure fits with detailed observations of our galaxy, the Milky Way, and other nearby galaxies is explained in the next chapter, and more detail on the composition of the arms and of the corresponding stellar population distribution is given. Here it only needs to be noted that the arms are fed by a stream of gas/plasma (comprising H, He and a trace of Li) ejected from the belt which condenses into stars which then form the visible spiral arms.

7.3   The full dynamic

The construction of the model that will explain spiral structure starts here. The first step is to extend the analysis of Chapter 5 to obtain a full model for orbits in the galactic plane and not just a formula for the rotation curve of such an orbit. The analysis applies to any rotating mass, but the results are only significant for truly enormous masses such as the hypothesised central mass in a galaxy.

Equation 5.8 (in Chapter 5) gave a formula for the tangential velocity in an orbit. What is needed is a formula for the radial velocity (again in terms of r) and these two will describe the full dynamic in the equatorial plane, which can then be used to plot orbits.

Intuitively there are two radial "forces" on a particle: a centripetal force because of the attraction of the massive centre and a centrifugal force caused by rotation in excess of that due to inertial drag. Thus a formula for radial acceleration of the following form is expected

(7.1) $$\ddot{r} = \frac{v_{\text{inert}}^2}{r} - F(r)$$

where $v_{\text{inert}} = v - \omega r$ and $F(r)$ is the effective central "force" at radius r, per unit mass. The same notation as in Chapter 5 is used here and in particular $\omega = \omega(r)$ is the inertial drag at radius r.

This will be proved in a similar way to the proof of conservation of angular momentum given in Chapter 5, using a geometrical argument which is valid in the inertial drag metric.

The idea is the same, namely to replace the central force by a series of central impulses at equally spaced small intervals δt of time and then take the limit as $\delta t \to 0$. Start by setting ω equal to zero. Consider Figure 7.6 (a copy of Figure 5.3 with extra labels). Recall that the motion of a particle

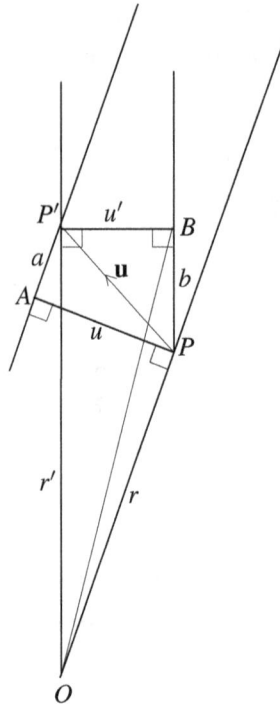

Figure 7.6: Diagram for radial acceleration

(of small unit mass) in the equatorial plane is being considered and that, at a particular time, it is at P and has just received a central impulse resulting in velocity $\mathbf{u}$. $a = |AP'|$ is the outward velocity (ie $\dot{r}$) at P (after the central impulse) and $b = |PB|$ is the outward velocity at P' before the central impulse. The effect of the central impulse is to subtract $F(r')\,\delta t$. Therefore if a' denotes the value of $\dot{r}$ at P' then $a' = b - F(r')\,\delta t$ or

$$(7.2) \qquad\qquad a - b = -F(r')\,\delta t - \delta a$$

where $a' = a + \delta a$. But by Pythagoras $a^2 + u^2 = ||\mathbf{u}||^2 = b^2 + (u')^2$ and hence

$$(a - b)(a + b) = \delta u\,(u + u')$$

where $\delta u = u' - u$ as before. Then substituting for $a - b$ from (7.2) gives

$$(7.3) \qquad\qquad (a + b)(-F(r')\,\delta t - \delta a) = (u + u')\,\delta u.$$

But recall from Equation (5.4) that $ur = u'r'$ which implies

(7.4) $$u\,\delta r + r\,\delta u = 0$$

to first order where $\delta r = r' - r$ as before. Now multiply (7.3) by r, reverse sign and substitute for $r\,\delta u$ from (7.4) to obtain:

(7.5) $$r(a + b)(\delta a + F(r')\,\delta t) = u(u + u')\,\delta r$$

But to first order $a + b = 2\delta r/\delta t$ (recall that a is $\dot{r}$), $F(r') = F(r)$ and $u + u' = 2u$. Thus (7.5) simplifies to

$$\frac{\delta a}{\delta t} + F(r) = \frac{u^2}{r}.$$

In the limit $\delta a/\delta t$ becomes $da/dt = d\dot{r}/dt = \ddot{r}$, which proves

(7.6) $$\ddot{r} = \frac{u^2}{r} - F(r).$$

Now reinstate ω. Exactly as in the previous proof, by locality the proof just given makes sense in the inertial frame at P in other words rotating with angular velocity $\omega = \omega(P)$. But $u = v - \omega r = v_{\text{inert}}$ and (7.1) is proved.

7.4 Computing radial velocity

Now specialise to the case $\omega = A/(r + K)$ (Equation 5.1) which was the formula for inertial drag coming from the Weak Sciama Principle. Here $A = K\omega_0$ and $K = kM$, where M, ω_0 are the mass and angular velocity of the central mass, and k is a weighting constant which can be taken to be 1 for purposes of exposition. The following formula for v (Equation 5.8) was found:

$$v = \frac{1}{r}\left(\int \frac{2Ar}{r + K}\,dr + C\right) = \frac{2A}{r}\left(\int 1 - \frac{K}{r + K}\,dr\right) + \frac{C}{r}$$

(7.7) $$= 2A - \frac{2AK}{r}\log\left(\frac{r}{K} + 1\right) + \frac{C}{r}$$

where C is a constant which can be read from the tangential velocity for small r. This implies:

(7.8) $$v_{\text{inert}} = 2A - \frac{2AK}{r}\log\left(\frac{r}{K} + 1\right) + \frac{C}{r} - \frac{Ar}{K + r}$$

Moreover for the purposes of investigation assume that $F(r)$ is the inverse square law $F(r) = M/r^2$. This is correct for the inertial drag metric based on Minkowski space (with Newtonian physics to first order) and is a good approximation for Schwarzschild and Schwarzschild–de Sitter provided r is not small. Thus:

$$\ddot{r} = \frac{v_{\text{inert}}^2}{r} - \frac{M}{r^2} = \frac{1}{r}\left[2A - \frac{2AK}{r}\log\left(\frac{r}{K}+1\right) + \frac{C}{r} - \frac{Ar}{K+r}\right]^2 - \frac{M}{r^2}$$

Multiplying by $\dot{r}$ and integrating wrt t (using a computer integration package) gives

(7.9)

$$\tfrac{1}{2}\dot{r}^2 = \int \ddot{r}\,dr = -\frac{C^2}{2r^2} + \frac{M - 2AC}{r} + \frac{A^2 K}{K+r} + A^2\log(K+r)$$

$$+ \frac{2AK(C + 2Ar)\log(1 + r/K) - (2AK\log(1 + r/K))^2}{r^2} + E$$

where E is another constant determined by the overall energy of the orbit. From this equation $\dot{r}$ can be read off (in terms of r). Moreover since there is a formula for v, there is also a formula for $\dot{\theta} = v/r$ (where notation has been changed to use the usual polar coordinates (r, θ) in the equatorial plane instead of (r, ϕ) as used in Chapter 5). From this it is possible express θ and t in terms of r as integrals. These integrals are not easy to express in terms of elementary functions but Mathematica is happy to integrate them numerically and this can be used to plot the orbits of particles ejected from the centre. Now use the hypothesis of Section 7.2 that the centre of a normal galaxy contains a belt structure, which emits jets of gas/plasma, which condense into stars. The orbits of these stars can be modelled and a "snapshot" of all the orbits taken at an instant of time, in other words a picture of the galaxy can be given. Excellent models for the observed spiral structure of normal spiral galaxies are found. This in done Section 7.6.

7.5   Simplified equations

There is a very convenient simplification for the equations given in the last section, which helps to explain how inertial drag controls the dynamic. For

Figure 7.7: NGC1365 and M51 images from NASA and Hubble site resp

most of an orbit in a galaxy $r \gg K$, since r varies up to 10^5 for the main visible disc whilst $K \approx M \approx .1$. This makes the fraction $A/(K + r)$ close to A/r and the formulae for v and v_{inert} reduce to $2A + C/r$ and $A + C/r$ respectively and then

$$\ddot{r} = \frac{A^2}{r} + \frac{AC - M}{r^2} + \frac{C^2}{r^3}.$$

There are good reasons for setting $C < 0$ (see Section 7.2) so that the AC/r^2 term acts to increase the gravitational pull. But the positive terms A^2/r and C^2/r^3 offset the central gravitational pull (the first for large r and the second for small r) and this allows long slow outward orbits which fill out the spiral arms.

7.6 Mathematica generated pictures

Below is the basic Mathematica notebook which generates galaxy pictures from the dynamics found in Section 7.4 above. The notation is as close as possible to the notation used before. A, K, r and v are A, K, r and v resp. E and C have been replaced by EE and CC because E and C are reserved symbols in Mathematica. M has been replaced by three constants Mcent, Mdisc and Mball. This is to allow an investigation of the effect of significant non-central mass on the dynamic. Mcent acts exactly as M above whilst Mdisc and Mball act as masses of a uniform disc or ball of radius rmax. Setting Mball = Mdisc = 0 reduces to the case of just central mass considered above. inert is v_{inert} and the other variables should be obvious from their names.

```
A = 0.0005; Mcent = .03; EE = -.00000345; CC = -10;
B = .00000015; Mball = 0; Mdisc = 0; K := Mcent;
rmin = 5000; rmax = 50000; iterate = 1000; step = (rmax -
    rmin)/(iterate - 1);
v := 2*A - 2*K*A*Log[1 + r/K]/r + CC/r;
    inert := v - A*r/(K + r);
Plot[{inert, v}, {r, rmin, rmax}, AxesOrigin --> {0, 0}]
rdoubledot := inert^2/r - Mcent/r^2 - Mdisc/rmax^2
                                     - Mball*r/rmax^3;
Plot[{rdoubledot}, {r, rmin, rmax}, AxesOrigin --> {0, 0}]
energy := -CC^2/(2*r^2) + (Mcent - 2*A*CC)/r - Mdisc*r/rmax^2 +
    Mball*r^2/(2*rmax^3) + A^2*K/(K + r) + A^2*Log[K + r] +
    2 A*K (CC + 2*A*r) Log[1 + r/K]/(r^2)
    - (2 A*K*Log[1 + r/K]/r)^2 + EE;
Plot[{energy}, {r, rmin, rmax}, AxesOrigin --> {0, 0}]
rdot := Sqrt[2*energy];
Plot[{rdot}, {r, rmin, rmax}, AxesOrigin --> {0, 0}]
ivalue := rmin + (i - 1)*step;
thetadot := v/r;
dthetabydr := thetadot/rdot ;
dtbydr := 1/rdot;
thetavalues = Table[NIntegrate[dthetabydr, {r, rmin,
    ivalue}], {i, iterate}]
tvalues = Table[NIntegrate[dtbydr, {r, rmin, ivalue}],
    {i, iterate}]
ListPolarPlot[{ Table[{thetavalues[[i]] - B*tvalues[[i]],
        ivalue}, {i, iterate}] ,
    Table[{thetavalues[[i]] - B*tvalues[[i]] + Pi, ivalue},
        {i, iterate}] }]
```

The program uses Equations 7.9 and 7.7 to express dt/dr and $d\theta/dr = \dot{\theta}/\dot{r} = v/(r\dot{r})$ in terms of r and then integrates numerically with respect to r in steps of size step. It then plots the resulting values of θ in the (r, θ) plane. If step is small this gives a good approximation to the orbit of a particle. To plot the spiral arms, it is necessary to allow the roots of the arms to precess. There is a new constant B which is the (apparent) rate of precession. Suppose that the roots are at radius r_0, where inertial drag is approximately A/r_0, then the inertial frame at that radius is rotating with respect to the background Minkowski metric with angular velocity approximately A/r_0, and, if the roots are stationary in that frame, they *appear* to be precessing with this angular velocity. B adds a linear term to θ to realise this. With B set to 0, the program sketches orbits. With B set nonzero the program sketches a snapshot of the spiral arms at a particular time. There might be some other effect causing precession and B can be adjusted to fit any such effect. In any case, it is necessary to guess r_0 in order to set $B = A/r_0$. Since the program starts at $r =$ rmin, a good first guess for B is $A/$rmin.

The program is intended for interactive use and the reader is recommended to investigate the output. Copies of the notebook with the settings used here can be collected from [4]. Note that these are all the same program; just the pre-set settings vary. Details of these settings are given in the descriptions which follow. For Figure 7.8 use Basic.nb and for Figures 7.9 (left and right) and 7.10 (left) use Full.nb. For Figure 7.10 (right) use Bar_galaxy.nb. Here are some hints on using it. As remarked earlier, the sketches are discrete plots obtained by repeated numerical integration. The number of plot points is set by iterate. Start investigating with iterate = 100 which executes fairly quickly and then set iterate = 1000 for good quality output. The plots are calculated in terms of r *not* time. The time values can be read from the tvalues table which is printed as part of the output. r varies in equal steps from rmin to rmax which need to be preset. You can't run to the natural limit for r (when $\dot{r} = 0$) but have to stop before this happens.

Set A to fit the desired asymptotic tangential velocity $2A$. For example to get $2A \approx 300$km/s set A = 0.0005. Set Mcent to the desired central mass. For example 10^{11} and 10^{12} solar masses are $M = .03$ and $.3$ respectively. Leave K set to equal Mcent unless you want to experiment with large values (which will increase the inertial drag effect for a fixed mass). Start with B set to $A/$rmin and adjust to get the desired spiral pitch. The integration constants C and E affect the picture mostly near the middle and outside resp. There are theoretical reasons for setting C to be negative because of the nature of the spiral arm generator (see Section 7.2) and with C set

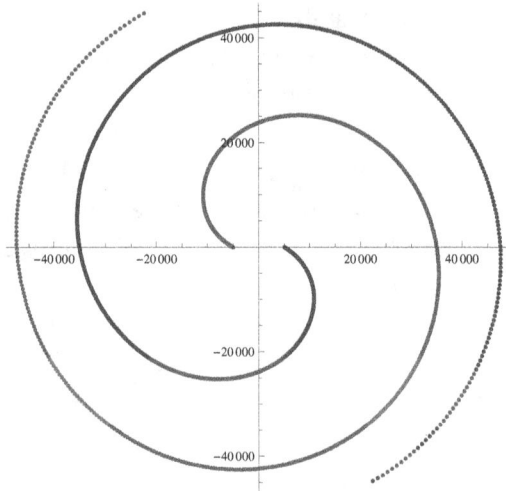

Figure 7.8: Output from the program as printed

negative, the roots of the spiral arms are offset in a way seen in many galaxy examples. E is a key setting as it determines the energy of orbits and hence the overall size of the galaxy. To get the most realistic pictures you need $\dot{r}$ to go to almost to zero at the maximum for r which you get by fine tuning E. To help with this tuning, the program plots the graphs of $v, v_{\text{inert}}, \ddot{r}$, energy and $\dot{r}$ so that you can adjust to get $\dot{r}$ and $\ddot{r}$ near zero at `rmax`.

Here now are some plots of orbits and galaxy arms obtained from this program. These should be compared with the images of real galaxies that are reproduced in (Figures 7.1, 7.3 and 7.7).

Figure 7.8 is the output from the program as printed above. M has been set to 10^{11} solar masses (all central) with tangential velocity asymptotic to 300km/s, `rmin` has been set to 5,000, B to A/rmin and `rmax` to 50,000 light years (corresponding to a visible diameter of 100,000 light years). Time elapsed along the visible arms is 5.5×10^7 years. The nature of the visible spiral arms will be discussed carefully in Chapter 8. Here merely note that the visible arms correspond to strong star-producing regions and bright short life stars, which burn out or explode in 10^5 to 10^7 years.

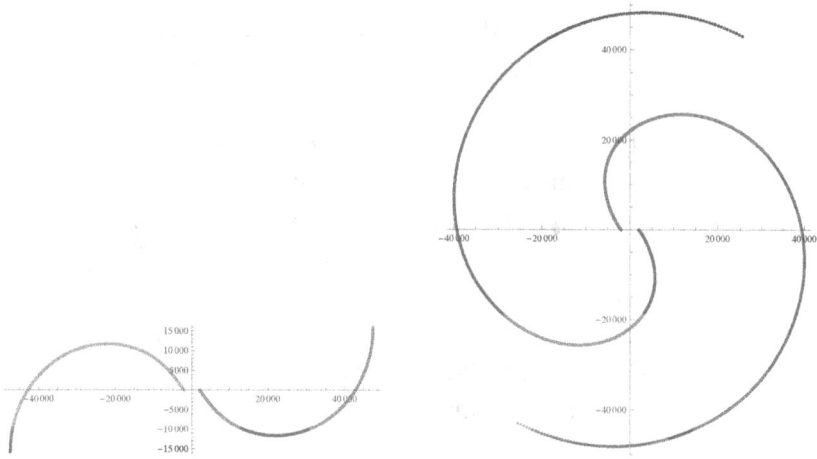

Figure 7.9: Left: orbits. Right: loose spiral

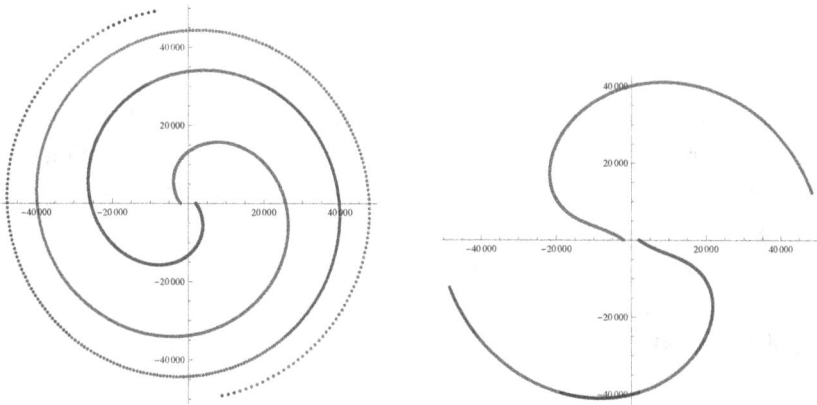

Figure 7.10: Left: tighter spiral. Right: bar.

Thus a total time elapsed of 5.5×10^7 years allows several generations of stars to be formed and to create the heavy elements necessary for planets such as the earth to be formed.

Figures 7.9 (left and right) and 7.10 (left) have the same settings with only B varied. The settings are similar to Figure 7.8 but with a small realistic contribution to the mass coming from Mdisc and Mball which are both set to 0.01 (1/3 of the central mass). rmin has been reduced to 2000 to get nearer to the centre. Elapsed time for all three is the same and again is 5.5×10^7 years. $B = 0$ for Figure 7.9 left, so these are actual orbits and B has been set to 10^{-7} and 2×10^{-7} resp for the other two to give a loose and a tighter spiral. Finally in Figure 7.10 right M has been reduced to 0.01 ($10^{10.5}$ solar masses) and the settings chosen ($C = -5$ and $B = 5 \times 10^{-8}$) to give a realistic bar galaxy. Elapsed time here is 10^8 years. Two classic examples of bar galaxies are NGC1300, Figure 7.3 (right), and NGC1365, Figure 7.7 (left). Both of these appear to contain two rather different structures: spiral arms and a superimposed dusty bar. The model given in Figure 7.10 (right) models an amalgam of these so there is a need for a better model for bar galaxies, and this is discussed in the next two subsections.

7.7 The bulge

In order to model bar galaxies more accurately it is necessary to consider another feature of galaxies which has only been mentioned in passing up to now, namely the central bulge. This is a chaotic collection of stars and other material lacking the dynamic coherence of the spiral arms. Most stars are on fairly tight orbits around the central black hole and there is a very large range of star types observed. There is a predominance of poulation II (cool red stars) and this accounts for the red colour of the bulge seen in many galaxy photos.

In terms of the accretion model constructed in Chapter 6 and Appendix C the bulge is analogous to the settling region where incoming matter loses its kinetic energy (KE) by interaction and settles towards the central black hole. The KE of incoming matter keeps up the energy levels and it is also fed from the central black hole in the same way as the generator. So there is analogous activity with small jets (not so organised as for the main jets that create spiral arms) and new star streams. The general appearance is of a cloud of stars which is usually spherical. But if affected by nearby strong gravitational fields it can take other shapes and this is precisely what happens in a bar galaxy.

Figure 7.11: NGC4394 (left) and archetypal bar galaxies (right)

7.8 Bar galaxies

Turning now to bar galaxies, here is a description from the Hubble site: "NGC 4394 is the archetypal barred spiral galaxy, with bright spiral arms emerging from the ends of a bar that cuts through the galaxy's central bulge", Figure 7.11 (left). The two pictures on the right of the figure are standard sketches of barred spirals based on this archetype but not on any real galaxy. Real galaxies never look like either of these! NGC 4394 itself has an extensive but rather chaotic spiral structure extending right into the central region marked by spiral lanes of dust. The arms are not well defined but what can be seen very clearly is that they do NOT emerge from the ends of the bar. Where the spiral arms emerge in a barred galaxy can be seen more clearly in the two classic bar galaxies pictured above, namely NGC1300, Figure 7.3 (right), and NGC1365, Figure 7.7 (left). In both cases the arms can be traced back to the centre as indicated in Figure 7.12.

Not all of the indicated arms are visible. For NGC1365 they are more-or-less visible all the way to the centre but for NGC1300, only a short section can be seen inside the radius of the bar. There are three possible explanations for this. Firstly the bar itself occludes the arms and secondly the nature of the arms (clouds of pre-stellar material condensing into violent star-producing regions) makes it likely that the arms are invisible near the roots; this is

Figure 7.12: NGC1300 (left) and NGC1365 (right) with spiral arms indicated

analogous to the way a gas flame burns: the stream of plasma needs to condense into clumps to form stars and ignite. The third possibility is that the generator has stopped emitting material for a while. Looking at a collection of galaxy photos it is possible to find clear examples where spiral arms stop and start, presumably because the generator runs out of "fuel" and needs to accrete some more.

The bar itself is not part of the spiral structure but a distorted central bulge, the distortion being due to the nearby massive streams of matter that are feeding the main spiral arms. In fact the bar and the arms are dynamically disjoint: orbits in the bar are local but elongated by the distorion from the pull of the arms; orbits in the arms are (like all arm orbits) long slow outward spirals lasting 10^8 years or more.

Chapter 8

Observations

Previous chapters have established a new model for spiral galaxies based on Sciama's principle. This has provided a satisfactory explanation for observed rotation curves without using "dark matter", and accurately modelled the spiral structure. The salient features of this new model are (1) a central rotating mass (presumably a black hole) of 10^{11} to 10^{14} solar masses which controls the dynamic via inertial drag effects and (2) a counter-rotating belt structure similar to the accretion disc around the (rather smaller) black holes in so-called "active" galaxies which feeds the roots of the spiral arms with pure H–He ions (with a trace of Li). The whole galaxy has a cyclical structure with matter ejected from the centre, condensing into stellar systems, moving outwards along the arms, burning out and falling back into the centre to be recycled. Thus the outward flow of gas mixes with dust and debris outside the belt and, as it flows outwards, condenses into violent star-producing regions illuminated by novae and supernova explosions. These regions sythesise the heavier elements needed for planets such as the earth to support chemically based life-forms. Solar systems containing such planets are formed further out along the arms.

This chapter considers detailed observations from our own galaxy which support this new model. More information on the entire structure will emerge in the process.

Topics covered are: early 21cm observations, Section 8.1, stellar populations, Section 8.2, the nature of SgrA*, Section 8.3, the position of the Sun, Section 8.4, globular clusters, Section 8.5, and, as a related appendix (Appendix D), an extended discussion of local stellar velocities. In particular, the new model provides natural explanations for the non-existence of hypothetical type III stars and for the peculiarities of the velocity ellipsoid for local stellar velocities.

8.1   21cm emission observations

The first comment is that the outward flow of gas along the arms of the Milky Way was clearly observed by Oort, Kerr and Westerhout [79] in 1958 and in subsequent surveys. The correct interpretation was made at the time but was later changed to attribute these observations to a hypothetical bar structure (for which there is little other evidence) see Binney and Merrifield [35, pages 17–18]. The idea is that the bar provides a massively asymmetrical central gravitating mass which allows for highly non-circular orbits, some parts of which fit the observed gas flows. This explanation is implausible for the same reason that the dark matter explanation of the rotation curve is implausible. In both cases a rotating dynamical system is proposed which is supposed to be stable but does not have a dominating central mass to provide stability.

It is worth remarking that a satisfactory model for bar galaxies is provided in Sections 7.7 and 7.8 and that in this model the bar itself is part of the central bulge structure, in which orbits are generally chaotic. It is also worth remarking that similar gas flows have been observed in other galaxies.

8.2   Stellar populations

In the model for galaxies proposed here, stars are formed by condensation in the outward flowing streams of gas coming from the central belt structure, loosely called "the generator". The outer layer of the belt is a plasma of H and He ions with traces of Li and other particles. Conditions here are similar to those hypothesised to have occurred just after the Big Bang and the consequent mix of light elements is the same. As modelled in Chapter 7, the outward flowing gas streams form into the familiar spiral arm structure. It is in these arms that stars condense. The residue of the gas streams, not condensed into stars, escapes the galaxy and feeds the intergalactic medium and this explains the observed proportion of light elements in the universe (which is one of the so-called "pillars of the Big Bang theory").

Near the roots of the arms, this condensation creates the observed violent star-producing regions with novae and supernovae. Here heavier elements are synthesised in abundance and moving outwards along the arms, the composition in the background gas stream alters to include dust and debris from this synthesisation and stars condensing further out have higher metalicity[1]. Thus for stars in the neighbourhood of the sun, fairly far out from the centre along an arm, there is a natural inverse correlation between the age of a star and its metalicity. Later a good estimate for the distance of the Sun from the centre of the galaxy will be found.

This is usually described in terms of "stellar populations": population II stars are older stars with low metalicity formed near the roots of the arm in which the sun lies whilst population I stars are younger stars formed further out, after enough population II stars have exploded as supernovae to provide the higher metalicty in these stars. The Sun is a population I star.

Under the Big Bang hypothesis, there should be a third population (population III stars) formed immediately after the Big Bang from pure H–He with zero metalicity. These stars have never been detected. In the model proposed in this book, they would have to be formed at the very roots of the arms. But, because of the cyclical nature of the model, outside the belt the galaxy is heavily polluted with dust and debris of various kinds coming from stellar systems falling back into the centre to be recycled. Thus the pure stream of H–He is quickly contaminated with traces of metals. Therefore stars formed even very near the roots will be contaminated with metals and be population II stars. Thus the model naturally explains the different stellar populations and why there are no population III stars observed. Notice that

---

[1]Metal is used here, with the misuse common in astronomy, to mean all elements heavier than He.

the difference between population I and population II stars is not their age, but where they are formed in the arms. Stars formed near the centre will be older by the time they reach the neighbourhood of the sun than stars formed further out. Thus for stars near the sun, there is a inverse correlation between metalicity and age, as is observed.

The model for star creation suggested here implies that stars like the sun formed away from the centre will typically be surrounded by planets condensed from the heavier lumps of debris in the vicinity. This suggests that solar systems like ours are the norm rather than the exception. This prediction has in fact already been verified by many recent observations.

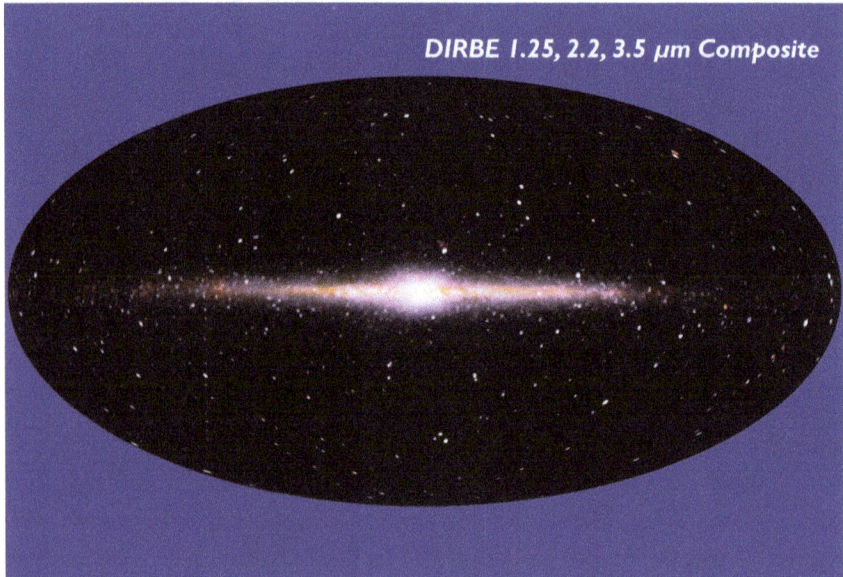

Figure 8.1: Composite image of our galaxy from the COBE satellite

8.3   Sagittarius A^*

There is a strong radio source at $SgrA^*$ which has been the subject of many observations. Observations of the proper motion of $SgrA^*$ using a very long baseline interferometer, due to Reid et al [83], suggested that the observed motion could be ascribed to the orbital motion of the Sun and that Sgr A* might in fact be at rest. Further the orbits of stars near Sgr A* have been carefully monitored over a period of twenty years or so. These observations establish that this object is massive (about 4.3×10^6 solar masses, presumably a black hole) and at a distance from the Sun of about 8.3kpc. For a good overview see Gillesen et al [50].

The suggestion by Reid et al that Sgr A* might be at rest with respect to the galaxy as a whole has led to the belief that it is in fact at the centre of the Milky Way and this has become now an accepted "fact" with Gillesen et al for example describing Sgr A* as "the Massive Black Hole in the Galactic Center". However it is not nearly massive enough to drive the dynamic of a full-size spiral galaxy, and therefore this conclusion directly contradicts one of the main hypotheses of this book. Thus it is necessary to advance another explanation for these observations.

Globular clusters have total mass varying up to around 10^7 solar masses and central black holes have been detected in many clusters. Moreover there is a well-established theory for mass concentration and black hole formation in clusters, see [23]. Indeed this is a natural phenomenon as clusters age. Stars will burn out and collapse and mass concentration will cause a group of collapsed stars to coalesce into a single black hole. The group of stars orbiting Sgr A*, together with Sgr A* itself have all the characteristics of a globular cluster near the end of its life with most of the mass coalesced into the central black hole and the remaining stars in orbit around the centre.

At this point the truly wonderful image of our galaxy, the Milky Way, Figure 8.1 obtained from data collected by the COBE satellite [2] should be considered. This image provides clear evidence that Sgr A* is not at the centre of the galaxy. The image uses Mollweide projection, which preserves area and central symmetry. Because of the conviction that Sgr A* is at the centre, this has been located dead centre in the image. The horizontal scale is galactic longitude covering the full $360°$ and it is linear. If Sgr A* was truly at the centre of the galaxy then this image would be symmetrical about both the central vertical and horizontal axes. It is clearly not. The bulge peaks rather to the left of centre and the main disc (seen edge-on) is also displaced to the left. Not quite so obvious, but also clearly visible, is vertical

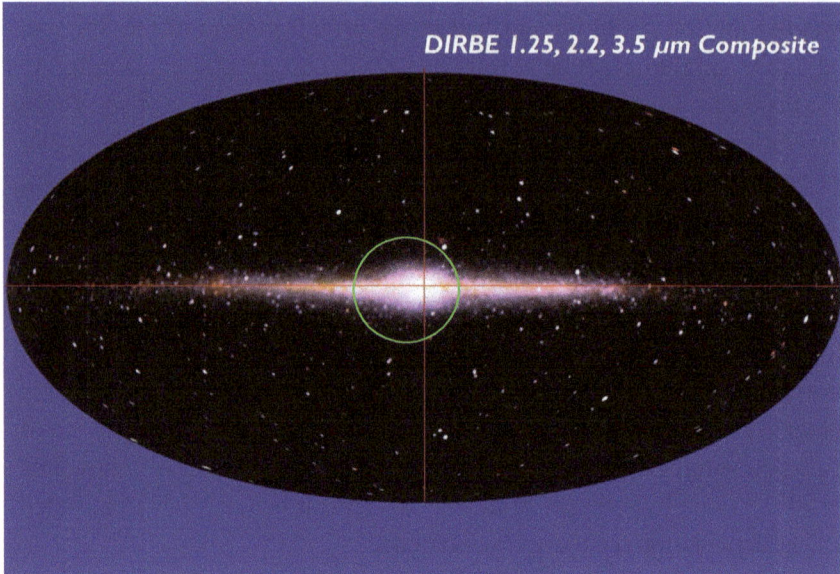

Figure 8.2: Image of our galaxy from the COBE satellite with axes and bulge highlighted

asymmetry, with the disc displaced slightly downwards from the central horizontal line. To help these asymmetries to be seen, the image has been reproduced in Figure 8.2 with the Mollweide axes superimposed in red and with a green circle centred on the widest part of the bulge. The bulge itself is not symmetric with a curious right hand smaller bulge superimposed. Below an explanation for this is suggested, and the smaller bulge has been ignored in placing the green circle. But even if it is not ignored and the two bulges are averaged, then it is still displaced to the left. (There is also a blown-up image of the centre of Figure 8.1 on the same site [3] with the asymmetry of the bulge and the vertical displacement both very clearly visible.) Because of the non-circular nature of spiral arms there is no reason to expect the main disc to appear symmetrical. But it is pretty symmetrical albeit displaced to the left. However vertical symmetry and symmetry in the central bulge is expected. The asymmetry corresponds to a displacement of SgrA* by 1° upwards from the true centre of the galaxy and between 3°

and $5°$ to the right (depending on whether the secondary bulge is ignored or the two are averaged).

So Sgr A* is not at the centre of the galaxy. Is there any reason to suppose that it is at rest? This assumption has become self-fulfilling with other velocities measured against it. If this assumption is dropped, there is no direct evidence to reinstate it. It would be necessary to measure average velocities for the galaxy as a whole, compensating for redshift due to a heavy centre (not Sgr A*) if any. There were early estimates using globular clusters (due to Shapley, see [35, page 8ff]), but no apparent recent global data to replace this assumption.

Now consider the bright region near Sgr A* which is associated with the smaller right-hand bulge. Here is a suggestion for what this might be, which also leads to a suggestion to explain the stellar composition of the cluster. It is suggested that the image here is looking straight down part of the arm coming out of this side of the central bulge and seeing an amalgam of strong star-producing regions which accounts for the brightness of the radio image. This implies that the region containing Sgr A* is full of pre-stellar material (dust and light elements) out of which stellar systems are condensing. The cluster has moved into this region, which accounts for the strange stellar composition — predominantly Wolf-Rayet and Type O with a sprinkling of young stars (the so-called paradox of youth). The young stars could result from the capture of clouds of pre-stellar material which have condensed into stars. The Wolf-Rayet and Type O stars are heavy old stars consistent with extreme age for the cluster as a whole.

8.4 Where is the Sun?

Since Sgr A* is not the centre of the galaxy, there is no direct way to measure the distance of the Sun from the centre. There is however a good deal of indirect evidence which places it at 17kpc (5×10^4 in natural units) or more from the centre. It is necessary to consider what is actually seen when looking at a spiral galaxy. There are several images reproduced in Chapter 7 to look at (Figures 7.1, 7.3 and 7.7). In all cases it is clear that the visible spiral arms are characterised by intense star producing regions populated by massive short life stars and that a region of smaller older stars such as our immediate neighbourhood would very probably appear quite dark from a distance. So it is expected to be some way outside the main visible disc (which is typically about 10^5 in diam).

There is also the timescale to consider. The visible arms mostly comprise massive short life stars which burn out or explode in 10^5 to 10^7 years. This fits well with the models constructed in Chapter 7 where matter takes from 10^7 to 10^8 to cover the length of the arms from centre. This gives time for several generations of stars to be formed and to create the heavy elements for population I stars (like the Sun) to contain (not to mention the earth). The Sun is about 5×10^9 years old and probably formed about half way along one of the arms of the galaxy. By now it must have moved beyond the visible arms. It is worth commenting that the spirals found in the models constructed in Chapter 7 have very shallow pitch near the outside (where both $\dot{r}$ and $\ddot{r}$ are small) and the outward movement slows down very considerably there. This means that Sun may be just a short way outside the visible arms, more-or-less on the edge of the visible disc at about 5×10^4 out from the centre.

Finally there is conclusive evidence again from the COBE satellite image, Figure 8.1. The visible arms clearly lie to one side of the Sun. They thin down to almost nothing for about half (or a little more) of the full circle represented by the centre line on the diagram. This puts the Sun right on the edge the main disc, or just outside, at again about 5×10^4 from the centre.

Incidentally the estimates found here agree closely with those made by Harlow Shapley in around 1918 based on distances to globular clusters (see [35, page 8ff]). These were later revised downwards and it is tentatively suggested that there may have been a systematic error in these revisions.

8.5   Globular clusters

Globular clusters comprise mostly population II stars. So they are formed very close to the central part of the galaxy. It is suggested that the instability in the central region, fed directly by energy from the black hole, occasionally throws a huge flare of gas (the usual H–He mixture) in a direction other than in the galactic plane. This could happen as a short-life "storm" structure. An analogy would be a cyclone forming in the earth's atmosphere. Such a flare could condense to form a tight cluster of population II stars: a globular cluster in fact.

There are about 200 globular clusters in a galaxy and they have lifetimes of 10^{10} years or more so, to maintain the population, there need be only one new cluster formed every 10^7–10^8 years. Thus this model makes it possible that the constitution of a galaxy might be more-or-less constant over a timescale several orders of magnitude greater then current estimates. In the next chapter these ideas are pursued and their consequences for global cosmology discussed.

8.6 Local stellar velocities

There has been a huge effort expended mapping the velocities of stars in the neighbourhood of the sun. There are some paradoxical properties of these excellent observations. In particular, the symmetries in velocity variations that would be expected from the current dynamical model of the galaxy (with stars moving in circular orbits) are not observed. The "velocity ellipsoid" which expresses this variation does not have the line from the Sun to the galactic centre as a principal axis, as would be expected from symmetry; the deviation of these two directions is called "vertex deviation". Further, vertex deviation varies systematically with stellar age. The dynamical model proposed in this book has no such symmetry and these paradoxical aspects disappear. Further vertex deviation and its correlation with age have very natural explanations.

The discussion is fairly technical and has been postponed to an appendix (Appendix D).

Chapter 9

Cosmology

This chapter discusses cosmological consequences of the model for galactic dynamics constructed in the earlier chapters and starts by considering the Big Bang theory. No abstract theory in physics has ever captured the general imagination in the way that this theory has. Even the fine detail has passed into everyday usage. Here for example is an excerpt from a review from *The Guardian*:

> *The modern hunger to accord food spiritual "meaning" seems a relatively recent development: it is refreshing to note the absence of such inflated claims, for example, in the much-loved 1931 American cookbook The Joy of Cooking, by Irma Rombauer.* [Description of low-key rhetoric in this book omitted]
> *Yet since then foodist rhetoric has, like the early universe, experienced a period of rapid inflation. The foodist movement is desperate to claim other cultural domains as inherent virtues of food itself, so as not ever to have to stop thinking about stuffing its face. Food becomes not only spiritual nourishment but art, sex, ecology, history, fashion and ethics...*
> Extracted from: The Guardian 29 Sep 2012, Review section, Steven Poole "Get stuffed"

Given such universal appreciation of the fine detail of the theory, it seems churlish to prove that it is wrong. But unfortunately this is the case.

After dismissing the Big Bang theory, the three so-called pillars of the theory: the distribution of light elements, the cosmic microwave background (CMB) and redshift are discussed. As has been mentioned before, the observed distribution of light elements is accounted for using the proposed central generator for spiral arms in Chapter 7, see Section 9.2. The other two pillars (redshift, Section 9.4 and the CMB, Section 9.5) use the new model of the universe proposed in this book, see Section 9.3. Other topics discussed are gamma ray bursts, Section 9.3 and Section 9.6, the origin of life, Section 9.7 and an extended discussion of the quasar–galaxy spectrum, Section 9.8.

9.1   The Big Bang?

It has been observed several times that the model of galaxies that is proposed could be stable over a huge timescale (perhaps 10^{16} years or more). There is a natural cycle with matter ejected from the centre condensing into star populations with metalicity increasing with distance from the centre. Stars move out along the visible spiral arms and burn out before gravitating back towards the centre to be recycled. The contrary hypothesis, that the galaxy is only just older than the oldest known stars (or not quite as old as the oldest known globular clusters — see below) is just about possible, but is not credible in the light of galactic observations. There is a continued vigour to the star producing regions visible in all galaxies, which suggests a steady renewal of material from the centre and a long-term steady state.

Furthermore there are now several pieces of direct evidence that the Big Bang hypothesis is wrong. Globular clusters have just been mentioned. Although this is quite an old piece of evidence it is nevertheless completely solid; widely ignored, it effectively subverts the Big Bang theory.

Stellar evolution theory is very well-established, having an excellent fit with a huge body of observations; evolution for globular clusters is based firmly on stellar evolution. It is as solid a theory as any theory in physics. There are globular clusters in *this galaxy* which are 15 billion years old or more. This means that the galaxy itself must have been around for a good while longer than that. The Big Bang happened 13.7 billion years ago. There is a rather amusing chapter in Binney and Merrifield [35, Chapter 6] about this.[1]

To add to this, there are several recent observations of what should be features of the early universe, showing for example galaxies in the formative stages, which stubbornly refuse to show anything other than normal galaxies that might be seen nearby. The first clue that something was very much amiss, was provided by the space-based Hubble telescope. In 2003 the Hubble telescope was pointed at a dark part of sky, where it is possible to see back to near the Big Bang, and left running for a long time. The resulting "Hubble ultra-deep field" (HUDF) is published on the web [8]. It contains a wealth of information, some of which is so important for the arguments in this book, that Appendix E has been devoted to its properties. In the HUDF there are clear full-size galaxies which are so small (and therefore remote) that they are far too close to the Big Bang to have developed. A typical example is the very distant spiral galaxy (VDSG).

To follow arguments about the HUDF here and in Appendix E, the reader is recommended to download a copy of the highest resolution jpeg of the HUDF as instructed in the bibliography at [8]. To help find a particular galaxy or image, intrinsic coordinates are given from the bottom left, where the height and width are 1 unit and coordinates are taken mod 1 (so that a negative number is a coordinate from the right or top). The VDSG is at (.40, .26). A snippet of the field with this galaxy in it is reproduced as Figure 9.1 (left).

The image is somewhat distorted and this is a characteristic feature of the HUDF which will be explored in Appendix E. In brief, there is a background noise of gravitational waves which causes optical distortion. But assuming that this galaxy is what it appears to be, a full-size spiral of say 10^5 light years diameter, then by measuring the image and knowing that the HUDF has a linear size of 2.4 arc minutes, it can be calculated that this galaxy is 11×10^9 light years away, and was fully formed a mere 2.7×10^9 years after

---

[1]There are some recent attempts to square this circle (see for example [21]) but they feel like fudges to the author.

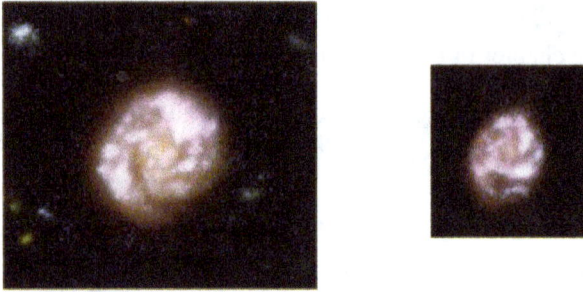

Figure 9.1: The very distant spiral galaxy (left) and a possible smaller example (right)

the Big Bang. This is far far too early for a full-size galaxy to have formed under standard theory. There is another even smaller example at (.17, .36) Figure 9.1 (right). This image is probably too distorted to definitely label as a full-size spiral, but if it is assumed that this too is a galaxy of 10^5 light years diameter, then it is 13×10^9 light years away and formed within 700 hundred thousand years after the Big Bang!

The discovery of distant far-too-large spirals near the Big Bang has been confirmed several times using different observations. Here for example is a news item from Physics.org [12] about Abel 383 a gravitational lensing image from the Hubble site [7]:

April 12 2011: First galaxies were born much earlier than expected

The giant cluster of elliptical galaxies in the centre of this image contains so much dark matter mass that its gravity bends light. This means that for very distant galaxies in the background, the cluster's gravitational field acts as a sort of magnifying glass, bending and concentrating the distant object's light towards Hubble. These gravitational lenses are one tool astronomers can use to extend Hubble's vision beyond what it would normally be capable of observing. Using Abell 383, a team of astronomers have identified and studied a galaxy so far away we see it as it was less than a billion years after the Big Bang. Viewing this galaxy through the gravitational lens meant that the scientists were able to discern many intriguing features that would otherwise have remained hidden, including that its stars were unexpectedly old for a galaxy this close in time to the beginning of

the Universe. This has profound implications for our understanding
of how and when the first galaxies formed, and how the diffuse fog
of neutral hydrogen that filled the early Universe was cleared.
Credit: NASA, ESA, J Richard (CRAL) and J-P Kneib (LAM)
Acknowledgement: Marc Postman (STScI)

And another from Nature (1 April 2009) [10]:

News: Early galaxies surprise with size
Astronomers revise galaxy-formation models with the discovery
that early galaxies could have grown fat — fast.
Eric Hand
Slurping up cold streams of star fuel, some of the Universe's first
galaxies got fat quickly, new observations suggest. The findings
could overturn existing models for the formation and evolution of
galaxies that predict their slow and steady growth through mergers.
Researchers using the Subaru telescope in Hawaii have identified
five distant galaxy clusters that formed five billion years after the
Big Bang. They calculated the mass of the biggest galaxy in each of
the clusters and found, to their surprise, that the ancient galaxies
were roughly as big as the biggest galaxies in equivalent clusters in
today's Universe.
The ancient galaxies should have been much smaller, at only a fifth
of today's mass, based on galaxy-formation models that predict slow,
protracted growth. "That was the reason for the surprise — that it
disagrees so radically with what the predictions told us we should be
seeing," says Chris Collins of Liverpool John Moores University in
Birkenhead, UK. Collins and his colleagues publish the work today
in Nature [38].

The quote has been curtailed. The rest is about patching up the theory. It
is necessary to be blunt about all these observations. They show that the
Big Bang theory is *wrong*. Of course, because so much has been invested in
the theory, no-one has admitted that it is wrong and indeed a strong fiction
is being maintained that it is being corrected. This is not going well. For
example here is an excerpt from the abstract for a cutting-edge seminar
given at Warwick on 22 May 2013:

Once considered the simplest class of galaxy to model and explain,
the assembly history of early type galaxies still presents many

puzzles. Spectroscopic observations show that the most massive
examples completed their star formation earlier than that in their less
massive counterparts, in apparent contradiction to popularly-held
hierarchical models.

What is being said is that larger galaxies were formed earlier, which is
obvious if there is no time zero to contend with, but which causes serious
problems when there is a time zero and the galaxy formation is far too close
to it! The Big Bang theory has a strong analogy with the flat earth theory.
In terms of this analogy, in these observations of very distant full-size spiral
galaxies, cosmologists are looking directly at the horizon and watching
ships sailing over it and still insisting that there is nothing beyond it.

The crisis in physics described on page 138 provides another proof that the
Big Bang theory is wrong. Calculations of the HLSW constant from local
observations (supernovae observations) and, assuming the Big Bang, from
the early universe (origin of the CMB) give incompatible figures within
observational error. The resolution is that the Big Bang model for the
CMB is wrong. Another model for the CMB which is compatible with the
supernovae data is given in Section 9.5 below.

So the Big Bang hypothesis is wrong and alternative explanations are needed
for the evidence that currently supports it. There are three so-called "pillars"
of the Big Bang theory: redshift, the distribution of light elements in the
universe and the cosmic microwave background. Explanations for all three
are given in the next four sections.

9.2   The distribution of light elements

The explanation for one of the pillars, the distribution of light elements in
the universe, has already been anticipated in outline in Section 8.2. To recap,
recall from Chapter 7 that near the hypermassive black hole in the centre of
a spiral galaxy is an accretion structure called the "belt" or "generator". It
is extremely hot, being fed energy both by accretion and by gravitational
induction from the black hole, and this causes a plasma of quarks to form
near the black hole. Moving outward from the centre the temperature drops

until ordinary ionised matter starts to form. This is exactly what happens in the standard Big Bang model, except that it takes place over space and not time. This produces the same mix of elements as in the Big Bang (H and He and a trace of Li and other particles, with ratio by weight of H and He roughly 3:1). There is a level that is equivalent to the last scattering surface in Big Bang theory where energy is radiated outwards. (An aside: this seems to be analogous to the Eddington sphere in the quasar model of Chapter 6.) Again as explained in Chapter 7 the belt also emits the streams of matter that feed the roots of the spiral arms (with the same mix of light elements) and the residue of these streams, not condensed into stars, escapes the galaxy and feeds the intergalactic medium and this explains the observed proportion of these elements which is the second of the three pillars. Notice that the universe as a whole is also cyclic with galaxies feeding the intergalactic medium and also being fed by accretion from this medium.

9.3 De Sitter space

To explain the other two pillars it is necessary to model the universe as a whole. The model is based on de Sitter space, the Hoyle Universe mentioned in Section 3.2. Appendix B is devoted to de Sitter space, where properties, such as the fact, stated in Section 3.2, that all time-like geodesics are equivalent, are proved, see Proposition B.1. The most elegant description of de Sitter space (dentoted deS) is due to Klein. It is the analogue of a sphere in Minkowski space (the simplest space in which relativity takes place) in other words the set of points (events) at a fixed distance from the origin in *Minkowski 5-space* — ordinary 5-space with the Minkowski metric:

$$ds^2 = -dt^2 + dw^2 + dx^2 + dy^2 + dz^2$$

This enables us to picture deS as a hyperboloid of revolution as drawn in Figure 9.2 for Minkowski 3-space. This gives an accurate represetation of the points (or events) in deS but a very misleading idea of the metric. As in special relativity, there are motions, *hyperbolic rotations* or *shears* that move points along rectangular hyperbolas (see Figure A.7). There is a shear

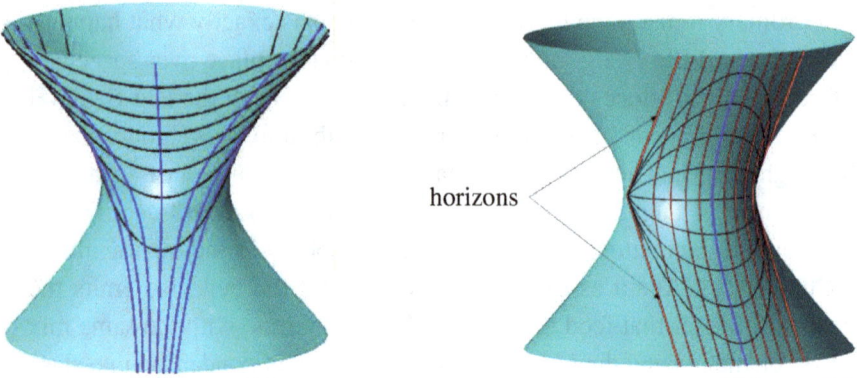

Figure 9.2: Figures reproduced from [77] of de Sitter space showing the central expanding frame (left) and the de Sitter static frame (right)

carrying the central horizontal circle into any ellipse given by intersecting with a plane through the origin, so all of these are equivalent. The picture does give a good idea of the linear structure of deS, geodesics are given by planes through the origin, geodesic 2-spaces by 3-spaces through the origin etc. The left picture in Figure 9.2 shows the expanding frame based on the central (home) geodesic (blue) with transverse flat 3-spaces (1-spaces in the picture). Note that time and expansion are upwards in the picture. The right picture shows the de Sitter static frame which has horizontal geodesics (black ellipses) and transverse time-like flow lines (orange) only one of which (central blue line) is a geodesic. Thus although this is a static frame, the time-like flow is highly unnatural. This was an unhappy first choice of coordinates for the space and obscured its perfect symmetry (like the 4-sphere it has a 10 dimensional symmetry group).

Light paths in Minkowski space are straight lines at $45°$ to the vertical and light paths in de Sitter space are those light paths in Minkowski space that lie in the hyperboloid, see Figure 9.3, which illustrates a light cone in Minkowski space meeting de Sitter space in a light cone. Thinking projectively, a forward light line is a tangent to the sphere S_+ at plus infinity (at the top) and a backward light line is a tangent to the sphere S_- at minus infinity.

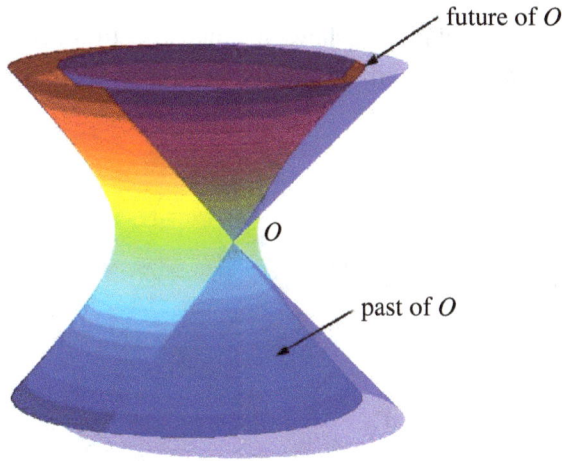

Figure 9.3: Light-cones: the forward light-cone in deS is the cone on a 0-sphere (two points) in the dimension illustrated (ditto backwards), in fact each is the cone on a 2-sphere. The figure is reproduced from [77].

Expansion

De Sitter space has a very suggestive property: any two geodesics which are not cofinal (ie not converging to the same point at $+\infty$) always eventually move apart in positive time at an increasing rate which rapidly tends to exponential separation. This is exactly what happens in the current "standard model" where *all* geodesics allowed by Weyl's postulate move apart at all times. But the standard model starts with a Big Bang singularity and is very unsymmetric. Thus although perfectly symmetrical, and having no Big Bang singularity, expansion is built into de Sitter space. There is a very interesting fact that follows from this: consider two observers A and B moving on different geodesics which are not cofinal. Since they eventually move apart faster than the speed of light, communication becomes impossible. Indeed there is a definite finite time b^* for observer B after which a light path from B cannot reach A. But A does not see this happening: A sees B for ever in A's time. B appears to be moving away faster and faster getting more and more redshifted. The moment b^* when B goes out of contact appears to A to be at time plus infinity. This effect is illustrated in Figure 9.4 on the right. The left-hand picture shows the dual effect for geodesics coming into contact and will be described shortly.

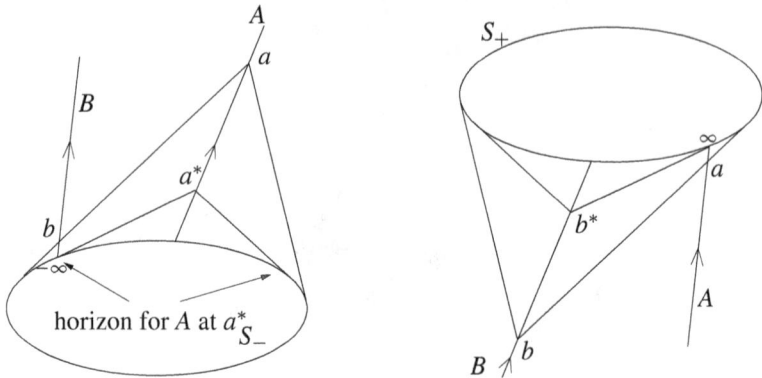

Figure 9.4: Left: first encounter   Right: last contact

The pictures in Figure 9.4 are projective pictures in which the light spheres, which are at infinity in deS, are represented by finite spheres, and geodesics are represented by straight lines, with null geodesics (light lines) tangent to the light spheres. S_+ is the forward light sphere and S_- the backwards light sphere. One can think of these pictures as obtained by projecting from the origin in Minkowski space onto a hyperplane not through the origin. In the pictures a typical light path from B to A leaving B at time b and arriving at A at time a is shown. In the right-hand picture the time b^* for B is the time when the light cone for B meets S_+ where A meets it, and after this the future for B does not contain any points of A and communication ceases. For A this happens at time $+\infty$.

Contraction

The perfect symmetry of deS implies that there is a dual effect for backward time. Backwards geodesics also separate exponentially and there is a contracting frame based on any time-like geodesic comprising geodesics which converge to the given geodesic in forward time. For a picture, take the left picture in Figure 9.2 and turn it top to bottom to get the contracting frame based on the home geodesic. Thus contraction is also built into de Sitter space. The way that the expanding and contracting frames fit together is similar to the way that expansion and contraction fit together in the hyperbolic (non-Euclidean) plane, see eg Figure 9.5.

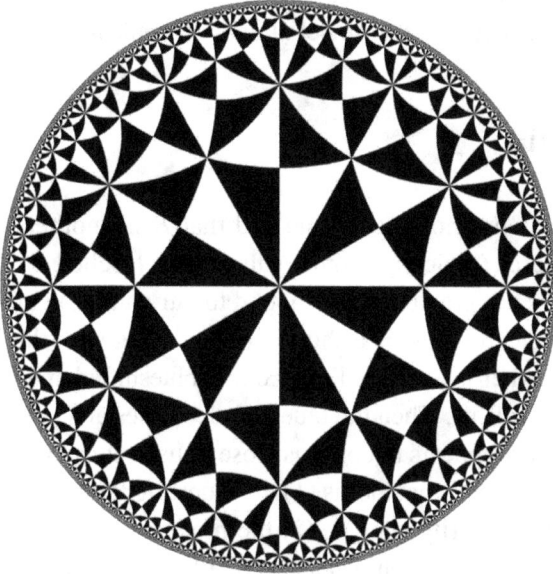

Figure 9.5: (6,4,2)-tesselation of the non-Euclidean plane from [18]

The dual effect to the "last contact" shown in Figure 9.4 (right) is the "first encounter" illustrated on the left. The time a^* for A is where the backwards light cone for A crosses over B at time $-\infty$ for B. In a very short period of A's time, B bursts over A's horizon and its entire history back to $-\infty$ is seen by A. The dual of the infinite redshift at time ∞ for A is an infinite blueshift at time $-\infty$ for B and the energy received by A is a burst of energy starting in the high gamma ray band and rapidly dropping to ordinary light and radio waves.[2] This is qualitatively similar to observed gamma ray bursts (GRBs) and MacKay–Rourke [72] (see also Appendix F) propose this as a possible explanation for these, and the reader is directed to this paper and appendix for fuller details and pictures. The theory fits a lot of the facts. But notice that if B radiates uniformly then the energy received by A in the

---

[2]This can be regarded as an extreme example of "relativistic beaming".

burst is infinite. So we need to assume that B only radiates for a finite time, or some other regularity hypothesis, to reduce the energy received to the finite bursts observed. Shortly another way to regularise will be proposed as a consequence of the model for the cosmic microwave background (CMB). But first it can be seen that the model already fits redshift observations.

9.4   Redshift

Assume that A is our home geodesic and that B is another geodesic on which a light source (typically another galaxy, also labelled B) is travelling. Assume that B is not gravitationally bound to our galaxy, ie not part of the local group. Typically there is a short period starting with the gamma ray burst at time a^* when we see the light from B blueshifted and B appears to be travelling towards us. Then the blueshift decreases and gets replaced by redshift: the galaxy appears to come as close to us as it can and then moves steadily away. The key behaviour is (1) the blueshift period is typically very short (it can be arbitrarily short)[3], (2) the movement away rapidly tends towards exponential separation (ie an exact HLSW law[4]) and (3) we see this phase for the entire remaining lifetime of the universe, which in de Sitter space is for an infinite time. There are graphs illustrating this in Figure F.3 and [72, 2.5, 2.6].

Now assume a uniform distribution of light sources radiating to us from other geodesics. All but a tiny number appear to be travelling away from us and to fit closely to an exact HLSW law, because we see the ones travelling away for an infinite time and the ones coming towards us for a short time. Thus almost all (in a measure-theoretic sense) are close to an exact HLSW law. It is natural to ignore the few exceptions as outliers and decide that *all* light sources are on the HLSW law and *the whole universe is expanding*. It is an extreme example of "Observer Selection Bias". We ignore the evidence that the universe is also contracting because it is so slight in comparison. Indeed most of the energy from incoming sources (the contracting bit) arrives in the

---

[3]See page 232, short blueshift periods t_B are distributed like t_B^{-5} and therefore skewed towards very short periods.
[4]See the convention on page 21.

form of GRBs and, as will be seen shortly, as the energy that is thermalised and seen by us as the cosmic microwave background. Neither of these is thought to be anything to do with expansion or contraction. In fact the universe is in a steady state and the expansion is a grand illusion.

Unlike the Bondi–Gold–Hoyle SST, the incoming matter that balances expansion is not created; it does not appear from nowhere. It is in the universe all along. It is just that our vision is incomplete. There is a similar incompleteness for forward time in both de Sitter space and the standard Big Bang model. Although a galaxy appears to us to be there for ever and just getting further away and more red-shifted, in fact it goes out of contact with us in a finite time it its own frame. This view makes it clear that the standard model is incomplete. It is only the forward half of the picture.

Author's remark   Observer selection bias as described above was proposed by Robert MacKay during the collaboration leading to [71] as a possible explanation for redshift. Although not playing a part in the paper as published, my view now is that it is the key idea in the correct explanation of redshift.

9.5   Cosmic microwave background

So far a model for the universe has been found that fits observed redshift and also GRBs. This section covers the CMB.

The CMB is a highly isotropic thermal radiation field that appears to emanate from every part of the sky. In particular it comes from the apparently dark background where there are no visible stars or galaxies. It is thermal to better than 1 part in 10^5 and has a temperature of between 2.725K and 2.726K. It is apparently anchored in the Machian rest frame determined by distant galaxies. Motion of the earth with respect to this frame (approx 371km/s towards Leo) can be detected accurately from dipole anisotropy. There are small fluctuations in temperature which, under standard Big Bang theory, come from quantum fluctuations in the inflation field hypothesised to have smoothed out the universe when it was very small.[5]

Although typically described as a weak radiation field, the CMB is about 45 times more energetic than the background starlight field, see Table 9.1.

Table 9.1: Rough estimates of energy and photon number densities of the extragalactic background at various frequencies intended for order of magnitude calculations, reproduced from [66, Table 9.1], data from Hauser and Dwek [51]

Waveband	Energy density of radiation (eVm^{-3})	Number density of photons (m^{-3})
Radio (300MHz)	10^{-2}	$\sim 10^4$
Cosmic Microwave Background	2.6×10^5	4×10^8
Infrared (140–1000μm)	4×10^3	3×10^5
UV-optical near IR (0.16–3.5μm)	$\sim 10^4$	$\sim 10^4$
X-ray ($\sim$10 keV)	20	3×10^{-3}
γ-ray ($\sim$1 MeV)	10	$\sim 10^{-5}$
γ-ray ($\geq$10 MeV)	0.5	$\sim 3 \times 10^{-8}$

The gravitational fog horizon: a limit to visibility

To understand the observed CMB it is necessary to use the fact that the universe is filled with a "noise" of low-level gravitational waves (arising from the inertial drag fields hypothesised in Chapter 4, the relative motions of the massive black holes at the centres of galaxies, and other sources) which can be observed in the systematic distortion in distant images seen by the Hubble space telescope.

This telescope is powerful enough to look back almost to the Big Bang or, since the Big Bang is not part of this story, to about $V = 13.7$ billion light years. [V stands for "visibility" and this terminology will be justified

---

[5]The author's opinion is that this part of the Big Bang story is fantasy physics on a par with Arp's explanations for redshift reduction with growth in quasars.

shortly.] In 2003 the Hubble telescope was pointed at a dark part of sky, where it is possible to see back nearly this far, and left running for a long time. The resulting "Hubble ultra-deep field" (HUDF) is published on the web and contains a wealth of information. The most important information for understanding the CMB is that space-time is not uniform at these scales. You can look at the HUDF and see optical distortion due to a low-level gravitational wave noise signal that fills the universe; see Appendix E for details.

This implies that light cannot travel more than a definite finite distance (about V) before it is diverted significantly from its original direction. There is an apparent boundary like the apparent boundary in a fog where light is comprehensively scattered. So there is a natural horizon where light particles are randomised in intensity and direction (and, as will be seen shortly, in frequency as well) by the gravitational fog. The CMB comes from there. The meteorological term "visibility" describes this horizon perfectly. It is the limit to visibility in the universe and this justifies the use of V for the distance to this horizon.

Observations of the CMB carefully correct for all visible galaxies and other radiation sources, and therefore select this apparent boundary. Light (as seen by us) travels in typically random ways near the visibility horizon. It is very unlikely to travel far in any one direction. It is, if you like, a random walk. Thus nearly all the light crossing the horizon from the far side, and heading towards us, has source within a region R of depth about $V/3$ behind the boundary. Now light passing through a gravitational wave field exchanges energy with the field. This is by a process similar to the Rees-Sciama effect (a special case of the Sachs-Wolfe effect [29]). The description given in [29] can be expressed in our context as follows. Think of a random gravitational wave field as a sequence of gravitational wells and hills which vary over time. A photon that goes down a gravity well and then emerges after the well has become shallower gains energy from the field and increases its frequency (and vice versa). And similarly a photon entering a hill expends energy but does not get all of it back if the hill becomes smaller before it exits. Thus light is randomised both in direction, intensity and frequency by the gravitational wave field and this is a perfect scenario for thermalisation.

The region R contains about 3×10^{10} galaxies all emitting roughly thermally at about 3000K and, after mixing and thermalisation, the radiation emitted from R in our direction is a near perfect black body spectrum again of temperature about 3000K.

Now the visibility horizon is fixed with respect to us (it is determined by the distance that we can see clearly) and therefore is part of our expanding frame, so the radiation is subject to cosmological redshift. The horizon is about V (13.7 billion ly) away and the effect of this redshift (a little over $z = 1000$) is to reduce the temperature to the observed CMB temperature of approximately 2.7K. This reduction is exactly the same as in the standard Big Bang model.

The CMB energy coming in this way from the starlight field near the visibility horizon is not energetic enough (by a factor of about 45, see Table 9.1) and to correct this it is necessary to take account of the radiation coming from incoming sources in the contracting frame. The bulk of this radiation is not subject to extreme blueshift and contributes, at about the same temperature, to that which is thermalised by the visibility horizon. This energy boost from incoming sources accounts for the high level of energy in the CMB compared to other extraglactic fields. One can think of this extra energy as providing a backlight for the horizon.

Remark Another interpretation of the visibility limit and horizon is provided by the paper of Ellis et al [41]. The backward light cone at any point in the universe develops a system of caustics caused by variable densities of matter in the universe. These caustics cause systematic distortion and ultimately a dense fractal pattern of distortion. This correlates with the gravitational fog description given here.

The HLSW constant: a crisis in Physics

In the current consensus theory there is a "crisis" because the two ways of determining the HLSW constant — from nearby observations (eg supernovae [16, 17]) and from distant observations (properties of the CMB) — give incompatible results within experimental error, see [14]. In the theory exposited in this book, there is no crisis. The local measurements correspond to global curvature of the universe determined by the cosmological constant whilst the distant observations correspond to the visibility horizon caused by background gravitational disturbances. Although roughly the same, this is just a coincidence and there is no theoretical reason to expect these two to be exactly the same.

Horizon effect and dipole anisotropy

The visibility horizon is indeed a "horizon": a virtual barrier caused by the behaviour of light. As such it depends on the observer. It is fixed with respect to the observer and therefore the apparent radiation (the CMB) should be perfectly isotropic (like black body Hawking radiation at the de Sitter horizon [46]). But in the overview (above) it was mentioned that there is a dipole anisotropy caused by the motion of the earth. The explanation for this apparent contradiction is that the radiation takes a very long time to travel from the horizon to us. About 13.7 billion years in fact. During this travel time, the motion of the earth can have (and obviously has) changed.

Quantum fluctuations?

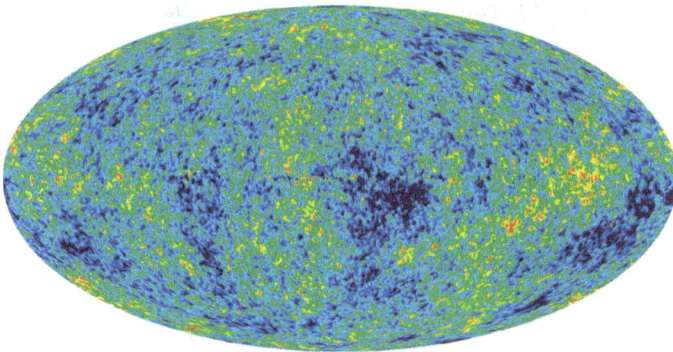

Figure 9.6: CMB temperature map

Now turn to the small fluctuations in power (or equivalently temperature) observed by many experiments and ascribed to quantum fluctuations in the inflation field as remarked earlier. These are illustrated in the CMB temperature map [1] reproduced in Figure 9.6. This map uses Mollweide projection and is an amalgam of nine years of data from the WMAP satellite. It shows the small variations in temperature of the CMB of the order of $20\,\mu$K. In this map, hotspots are red and yellow and cool areas green and blue.

The multipole power spectrum

The power spectrum is usually analysed using spherical harmonics and this results in the beautiful resonance curve reproduced in Figure 9.7. Here are some ideas for the origin of this curve. Longair has this to say (near the end of section 1.6 on page 22):

"Equally impressive was the fact that the determination of the two-point correlation function for galaxies determined from the large-scale galaxy surveys now overlapped the corresponding angular scales in the Cosmic Microwave Background Radiation and that these were in excellent agreement. In the most recent analyses, evidence has been found for a maximum in the two-point correlation function for galaxies corresponding to the first peak in the power spectrum of perturbations in the Cosmic Microwave Background Radiation..."

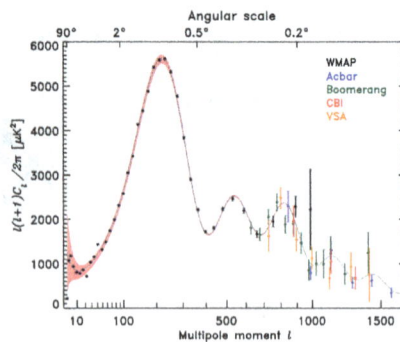

Figure 9.7: CMB Power Spectrum

The distribution of galaxies as seen by us has a great deal of regularity. For example $1°$ is a key angular measurement, correlated with the likely separation of nearby galaxies and it is this fact that Longair is referring to in the quotation above. Thus the anisotropies in the CMB are likely to be caused by the passage of the radiation through the galactic field on its way to us from the visibility horizon. This is a more natural explanation than the standard model which has both anisotropies (in the galactic field and the CMB) caused by the same fantasy physical quantum fluctuations in the "inflationary field". This leaves open the problem of why the anisotropies exhibit an (aparently acoustic) resonance curve Figure 9.7. There must be something in the large scale geometry of the universe causing this. But it causes the distribution of galaxies to have this partial regularity NOT the CMB. We observe it in the CMB because it is modulated by the galactic field on its way to our receivers. Another possibility is that the inertial drag fields from all the rotating black holes at the centres of galaxies form a standing wave field that causes the regularity observed in the power spectrum.

The writer of the wiki page at [22] comments: "Although many different processes might produce the general form of a black body spectrum, no model other than the Big Bang has yet explained the fluctuations. As a result, most cosmologists consider the Big Bang model of the universe to be the best explanation for the CMB." The model proposed here explains these fluctuations. In order for the Big Bang to explain them, a physically implausible hypothesis is needed (inflation) which breaks nearly all the basic laws of physics; consequently it is the Big Bang explanation which lacks credibility.

9.6 Redshift and GRBs revisited

Now return to the explanations for redshift and for GRBs given earlier to see how the CMB explanation has affected them. Redshift is quickly dealt with. Because of the visibilty limit we only see the redshift phase for a *very long time* (about the current estimate for the age on the universe) and not for an *infinite* time. Thus it remains true that nearly all sources exhibit redshift close to an exact HLSW law and the explanation is unchanged.

Turning next to GRBs, the infinite blueshift and infinite received energy from an emitter which emits uniformly are both regularised by the fact that light cannot travel more than about V without being diverted. The simplest way to see what happens is to use the duality between expansion and contraction mentioned in Section 9.3. An outgoing source of light appears to to be travelling directly away from us and to reach the visibility horizon after about 13.7 billion years, and has redshift of about $z = 1000$ when it disappears from view. Dually an incoming source pointed directly at us comes bursting over our horizon with blueshift (corresponding to the same frequency ratio of about 1000 — not infinite) and appropriate power. This is still a very bright burst of energy (a temperature increase of a factor of 10^3 and power increase of 10^6) which fits observations well. For GRBs not directed exactly at us, the enhancements will be rather less.

Here is a quick stab at a calculation. An incoming source of real temperature T will cause a GRB starting at up to about 1000T. This is quite reasonable: a source of temp about 10^6 K (temp of sun) would appear to be about 10^9 K which corresponds (using a black body model) to a wavelength of 3×10^{-12} m or 10^{20} Hz which is right in the gamma ray band (10^{19} Hz and up).

Now not all sources of light that come over the de Sitter horizon are visible. To be observable they have to come over the visibility horizon as well. So there is a large source of light outside the visibility horizon which feeds this horizon with energy that is re-radiated to us as the CMB (which has about 45 times the energy of background light coming from all visible galaxies). Thus, as remarked earlier, the visibility horizon is "backlit" and it is this backlighting that provides the bulk of the energy in the CMB.

9.7   Origin of life

The model for a galaxy has a built-in cyclic nature with solar systems created by condensation in the arms out of a mixture of the clean gas stream from the centre and the dust and debris left from stellar explosions and present as background throughout the galaxy, then living their lives whilst moving out into the outer dark regions of the galaxy and finally gravitating back into the centre to be recycled.

The timescale is huge. Probably several orders of magnitude greater than current estimates of the age of the universe. This is plenty of time for life to have arisen many times over on suitable planets. When these planets are destroyed by tidal disruption as they fall into the centre or by breaking up in collision with other objects, many of the molecules will survive and become part of the background dust out of which new planets are made. Thus in a steady state planets will start out seeded with molecules (probably in the form of very hardy viruses) which will help to start life over again. Indeed standard selection processes over a galactic timescale will favour lifeforms which can arise easily from the debris left over from the destruction of their planetary homes. This might explain how life arose on earth rather more quickly than totally random processes can explain.

There is evidence for this in the long-chain hydrocarbon molecules that are found in meteorites and detected in the intergalactic medium. See also the papers of Hoyle and Wickramasinghe (for example [57]): the model proposed here is fully consistent with their ideas on the present cosmic origin of micro-organisms.

9.8 The quasar–galaxy spectrum

This section returns to perhaps the most important consequence of the ideas presented in this book and probably the best way to understand them. The new model for galaxies fits ordinary galaxies into a spectrum of black hole based phenomena which includes quasars and "active" galaxies. The spectrum is conveniently ordered by the mass of the central black hole. As a very rough guide (in solar masses) these range from 10^7 or less to 10^{14} or more as follows:

Quasars: from 10^7 or less to 10^9

"Active" galaxies: 10^9 to 10^{11} approximately

"Normal" galaxies: 10^{11} to 10^{14} or more

All are highly active.

This spectrum has been discussed several times through the book and in the remainder of the chapter the main features are recollected. The first important point was mentioned very early in the book: quasars typically exhibit very large intrinsic (gravitational) redshifts as seen in observations of Halton Arp and others. Before proceeding it is worth looking briefly at another proof based on observations of this fact.

The Hawkins paper

An independent proof of the existence of gravitational (intrinsic) redshift in quasars is provided by a paper of Hawkins [52], which sets out to prove that quasars show redshift without time dilation (an impossibility since redshift and time dilation are identical in relativity and indeed in any metrical space-time theory), but in fact decisively proves that much of the redshift observed in quasars is intrinsic. For full details here, see the paper [86] on the author's web page; what follows is a quick sketch of the arguments.

Hawkins examines a large pool of observations of quasars. As has been mentioned before, radiation from quasars typically varies in intensity periodically over macroscopic time intervals from days to years. He makes a very careful selection from the pool (some more detail on this will be given later) and uses some very sophisticated analysis (which seems sound) to find a collection of quasars for which the macroscopic intensity variation does not exhibit time dilation correlated correctly with the observed redshift; indeed for this selection, the high redshift and the low redshift bins exhibit on average *exactly the same time dilation*. For full details, see [52].

This result is not paradoxical. What it shows is that for (a large subset of) this selection of quasars the sources of

(a) the radiation   and   (b) the time variation

are not in the same place. To enable discussion let us call these the *generator* and the *modulator* respectively. For the Hawkins sample, these must be subject to different redshifts, either cosmological or gravitational or a combination, with the modulator having lower redshift.

There are two possibilities:

(A)   The intrinsic redshift arrangement

Both are part of the same object (the quasar) and therefore both at roughly the same distance from us. This implies that the larger redshift (affecting the generator) is partly gravitational due to a nearby mass and that the modulator is further from the large mass and subject to a lower gravitational redshift.

This arrangement is precisely how the three-author-model described in Chapter 6 and Appendix C works. The generator is the Eddington sphere dividing the inner optically thick region from the outer optically thin region. No direct radiation comes from inside the Eddington sphere. The outer region contains strata of gas or plasma and further out there may be dust or more solid objects, all of which will typically be trapped in orbit around the central mass. The radiation from the generator passes through the surrounding layers on its way to us; the observed variations are due to non uniformity in these layers, and are naturally periodic with the possibility of several different periods coming from different layers superimposed. This is what is observed. Furthermore there is direct evidence for these layers in the Lyman-alpha-forest that is observed for some high redshift quasars, see Section C.7.

(B)   The microlensing arrangement

It is clear that the modulator must be on the light path from the generator to us. It does not need to be directly associated with the generator, as in the instrinsic redshift arrangement discussed above, anywhere on the path will do, provided it lies in a region of lower cosmological redshift. One way variations in intensity could arise would be if the path were subject to variable gravitational lensing effects or passing through a region of variable density. Both of these phenomena are called *microlensing*. There are indeed cases where this is known to happen (see eg Schild et al [91]) and if this happened to a large proportion of quasars then it would also explain the Hawkins result.

But is this plausible? It is not the existence of microlensing that is in doubt but its pervasiveness. It would be necessary to assume that there is a microlensing region happening to lie on the light path from *most quasars to us* and *close to us* as well. This is highly implausible unless nearly all space acts a microlensing region, eg if it is filled with suitable gravitational waves. There is indeed evidence for a gravitational wave field affecting distant observations, see Appendix E, but if this background field were strong enough to account for observed quasar variation then everything distant would have similar patterns of variation and no such variation has been observed for distant galaxies.

The only other way this could work would be if quasars were defined by the existence of a suitable microlensing region on the path to us. In other words if quasars were in fact distorted images of distant galaxies. But this possibility is again implausible because quasars have quite different radiation characteristics which could not be disguised by microlensing. So although apparently suitable as an explanation for the Hawkins result, microlensing has to be discarded, and the only remaining possibility is that a proportion of quasars in the sample have intrinsic redshift.

A basic question now arises. For any random sample of objects in the universe (which for the puposes of this discussion is assumed to be the standard expanding universe of current cosmology) there should be a correlation between redshift and time dilation whatever the mechanism that produces these locally. This is because the more distant objects will have both higher redshift, with the addition of cosmological redshift, and higher time dilation for the same reason. Hawkins has managed to find a sample which does not have this property. Obviously he must have used a non-random selection criterion at some point. And indeed he has. In an attempt to avoid the effect of another well-known correlation, between magnitude and redshift in flux limited samples, he has limited his sample to a very small magnitude range namely between magnitudes -25.5 and -22.5. This narrow sample contains high redshift quasars which have low luminosity and are close to us, and low redshift quasars with high luminosity which are distant. The former, being close to us, are subject to small cosmological time dilation effects and the latter to large ones. Thus the redshift–time dilation relation is skewed against the natural cosmological relation by the presence of these quasars whose redshift–time dilation is opposite to the natural relation, and this accounts for the redshifts in the sample not having the expected correlation with time dilation.

In passing, it is worth remarking that the well-known correlation (between magnitude and redshift in flux limited samples) mentioned above is probably due to observer selection bias. Most quasars are probably based around quite small black holes and the nearby ones (ie the ones with greatest magnitudes) will be the easiest to detect. The flux limitation eliminates the nearby ones with low (intrinsic) redshift and very high magnitude. Thus in any given flux limited sample, the higher magnitude quasars are more likely to be the nearby ones with high (intrinsic) redshift.

Quasars and redshift

Now return to the main topic, namely the quasar-galaxy spectrum, starting with quasars.

Quasars have been covered thoroughly in Chapter 6 and the associated Appendix C. Briefly, black holes aka quasars accrete matter from the surrounding medium and grow in mass. The key surface is the Eddington sphere which is analogous to the photosphere of a star. Inside the Eddington sphere is the *active* region where radiation is produced by interaction between infalling particles. This region is optically thick and only the boundary (the Eddington sphere) is visible and is where the radiation that is received comes from. The Eddington sphere can be very close to the event horizon and consequently subject to an arbitrarily high gravitational redshift and this accounts for the observed high intrinsic redshifts in some quasars. Because of the attenuation effect on power output of gravitational redshift [a factor $(1 + z)^{-2}$], the effective power output from the quasar can be far lower than the Eddington limit. Thus small quasars have both high (insrinsic) redshifts and low luminosity. One known example here is Sagittarius A* which has luminosity only 10^{-8} of the Eddington limit and corresponding redshift of 10^4 (which incidentally is why this quasar was first detected as a radio source). But in general quasars of very high redshift are unlikely to be detected because of their low power output.

As the mass grows with accretion, the distance between the event horizon and the Eddington sphere increases and gravitational redshift decreases. At the same time the central black hole gets increasingly masked by the accreting matter and more difficult to detect and measure. A large black hole tends to accumulate a thick inner region which masks it from the outside and allows the redshift to be very small, and conversely a small black hole has only a thin inner region and a large redshift. This natural effect explains why the active nature of normal (spiral) galaxies has not been directly observed. In a full size galaxy the central black hole is effectively shielded from view and the visible matter near the centre (the bulge) is sufficiently remote from this black hole that the usual way of estimating the central mass, using the virial theorem, does not yield any information.

Thus the huge black holes which power the dynamics of spiral galaxies (see Chapter 7) have not been directly detected and this is why the false assumption that SgrA* is the central black hole for the Milky Way has not been questioned before.

Quasars and active galaxies

Moving back down the spectrum to quasars. The smooth accretion of matter that happens for small quasars breaks down as the size rises to about 10^9 solar masses. The outer settling region becomes increasingly chaotic and the smooth accretion of matter into the central black hole stops. Matter trapped near the black hole now has no option but to form a rotating structure (called an accretion disc) as hypothesised in mainstream quasar theory. This is the start of the "active galaxy" stage for which accretion discs and associated jets have been directly observed.

The accretion disc continues to grow as the mass of the quasar/galaxy continues to increase by accretion.

Active and spiral galaxies

Once sufficient matter is trapped in the rotating accretion disc it begins to collect a significant amount of angular momentum. Since the total angular momentum is small (just collected from the pool in the surrounding medium) the central black hole must rotate the other way to acheive a balance. So there is now a rotating black hole with an orbiting structure, which rotates the other way, and which is referred to here as the *belt* to emphasise its likely toroidal shape. For definiteness call the rotation of the inner black hole "positive" and that of the belt "negative".

Jets produced by the belt will cause negative angular momentum to be lost to the system and increase the main positive rotation. There is now a significant inertial drag effect from the rotating black hole which increases the effective energy in the belt which becomes increasingly hot. A stable pair of opposite jets form and feed the roots of the spiral arms which are now growing. The belt is now the *generator* for the spiral structure. Stars form in the spiral arms and the whole galaxy radiates into the surrounding space and thus a limit in size is reached when the radiation balances accretion. From the model constructed in Chapter 7 the limiting size seems to be around 10^{14} solar masses.

The predominant life-form of the universe

It has been seen that quasars, "active" galaxies and larger spiral galaxies are all based around black holes, and that there is a natural way to suppose that these objects evolve over an extremely long timescale with points of the spectrum representing different ages of the same class of objects. A black hole grows with time by absorbing matter falling into its gravitational domain and the corresponding object moves along the spectrum. Moreover observations of Halton Arp and others [31] suggest that quasar/galaxies have the basic property of a life-form: reproduction. Quasars are often closely connected with parent galaxies and the natural supposition is that they have been ejected from them, for example, as has been mentioned earlier, Figure 6.1 shows what could be a family grouping of two parent galaxies and two offspring quasars.

Arp's observations also show (intrinsic) redshift decreasing with age which is consistent with the model given in this book where the larger the central black hole, the smaller the redshift. (Arp suggests some outlandish theories to explain this observation, which are quite unnecessary.)

So a quasar starts life as a comparatively small black hole which grows heavier with age. When it reaches the mass of an active galaxy it starts to throw out small black holes (quasars). This is the reproductive stage. Later it grows into a full size spiral galaxy and reproduction stops. Presumably, if it could be recognised, there is a final senile stage when the black hole disconnects from our space and the associated galaxy radiates away.

Finally it is worth remarking that nothing whatever is known about the inner nature of so-called "black holes". There is no such thing in nature as a singularity; black hole is simply the name given to another state of matter about which nothing is yet known. There are some fascinating observations due to Schild et al [91] which hint at a specific inner structure and which may perhaps shed some light here. Or perhaps by observing galactic clusters carefully it may be possible to deduce some of the rules governing this new state of matter — perhaps to begin to build up a proper physics for black holes. One point that needs to be addressed is why galactic centres are not even more massive. Black holes can combine to become more massive. So perhaps there should have arisen a set of super size galaxies grazing on ordinary ones etc. This does not appear to have happened. Why? The reason may be the mechanism described in Chapter 7 which limits size by boiling off excess matter, or the mechanism may be more elementary. Black holes over a certain mass may simply be unstable and spontaneously break up.

The lords of the universe

The main part of the book (before the appendices) finishes with some wild speculations. It is natural to think of these black hole based phenomena as part of *our* universe, but now turn the whole discussion over and try to see the universe from the point of view of these, the real inhabitants and creators. For them this airy space full of stars and planets must seem just a dream compared with the solid reality of their being. From the coincidence observed by Sciama (Equation 4.2) the rough volume of space is a simple function of the mass of these black holes, as if space is a property of them. Further these are the heavy weights which cause the curvature of space-time that gives the illusion of expansion. Indeed they carry nearly all the mass of the universe.

So the natural view from their point of view is that space is a property of their being. They create space as we create our dreams.

PART 3

Appendices

Appendix A

Introduction to relativity

Special relativity

A.1 Causality

> *The Moving Finger writes; and, having writ,*
> *Moves on: nor all thy Piety nor Wit,*
> *Shall lure it back to cancel half a Line,*
> *Nor all thy Tears wash out a Word of it.*

This is not a book of philosophy nor of poetry. It is a book of geometry. So why has this appendix started with a famous philosophical poem? It is because this poem expresses with great clarity the idea of "causality" which is the basis of relativity, the natural geometry of the universe. The essence of this famous quatrain is that the past cannot be altered, cannot be affected by anything that comes after it. Nothing that happens in the present moment can affect any time other than the future.

The most basic concept of relativity is of an *event*. An event just means something that has a definite place *where* it happens and time *when* it happens. Causality is the relationship between two events that the first event might affect the second. The quatrain says that, for causality to hold, the first event must precede the other in time.

155

Here is a contemporary example of causality in action. On the 11th of September 2001, in an act of unprecedented evil, two enormous and immensely strong skyscrapers were intentionally demolished with large numbers of people trapped inside. This event has affected almost every aspect of the current political environment of the whole world. But this influence has only been felt *after* this event. At no time *before* this event was there the slightest foretaste of the consequent loss of freedom and demonisation of sections of our communities. Events only affect the future. This is what causality means.

Figure A.1 is a basic diagram of causality.

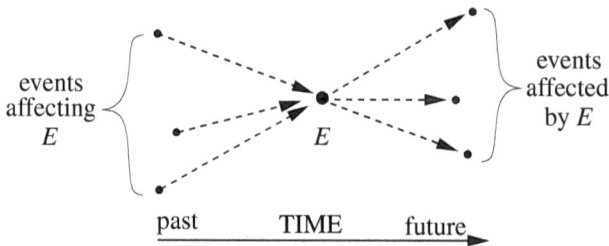

Figure A.1: Event E is affected by events in the past and affects other events in the future.

Relativists regard time as a dimension of exactly the same quality as space and they consider the two to be intermingled; they use the portmanteau word "space-time" for this intermingling. A space-time is a collection of events, ie points in space and time. Each event has both a time coordinate (*when* it happens) and a space coordinate (*where* it happens). In Figure A.1 you can think of the vertical axis (which hasn't been labelled) as space and then this diagram is a simple example of a "space-time diagram".

Another basic concept that will be used repeatedly is that of an *observer*. An observer just means the idealised path through space-time of a person. For each point in time the observer has a definite position in space, in other words for each time there is an event, namely that corresponding point in space-time. The collection of these points is called the *world-line* of the observer. At any point on this world-line there is one direction (along the world-line) which *appears* to be time and the perpendicular directions *appear* to be space. But this split into space and time depends on the world-line. A different observer will see a different split. This is a fundamental point of relativity:

Space and time are relative concepts which depend on the observer.

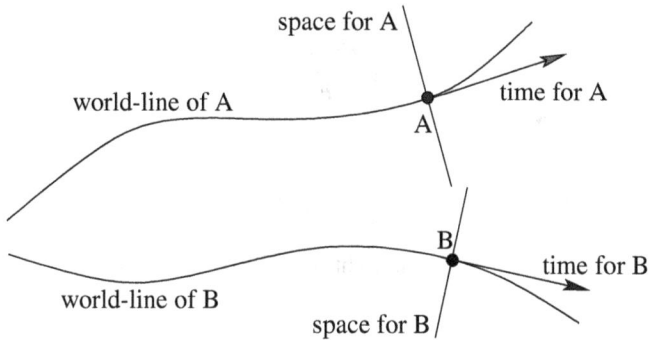

Figure A.2: Relativity of space and time: A and B are observers

A.2   The speed of light and Michelson–Morley

As well as the basic idea of causality, one key fact is needed: no information or effect of any kind can travel faster than the speed of light (about 3×10^8 metres per second). This limits the effect of a present event, not just to the future, but to those times and places in the future that can be reached at a speed up to the speed of light. So for two events to be causally related, it must be possible for a message originating at the first event to reach the second event at a speed less than or equal to the speed of light. A *light-line* is the world-line of a photon — a particle of light. In a space-time diagram a light line is a straight line. For each point of space-time and for each direction in space, there is a light line originating at that point going in that direction.

Figure A.1 has been updated in Figure A.3 with this new information. The set of events in the future which can be affected by E are bounded by the two outgoing light-lines from E (one going up and the other going down). The region comprising these events is called the *future of E*. Similarly the

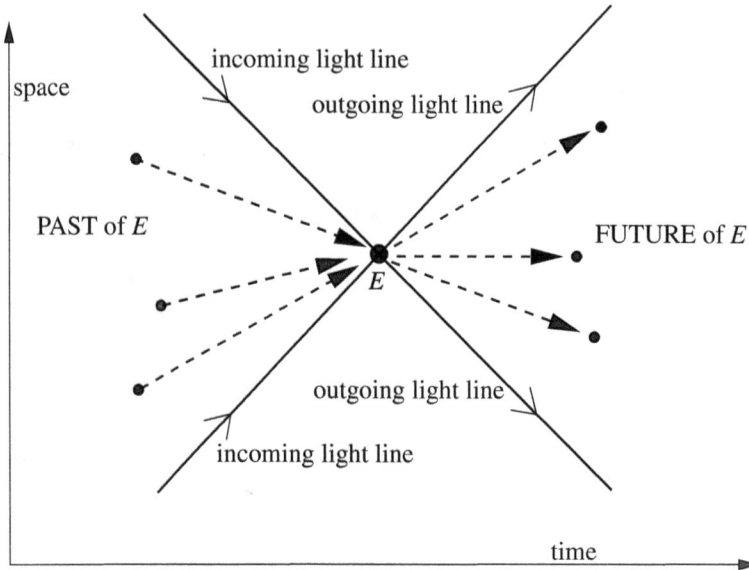

Figure A.3: Causality diagram with light lines. The top and bottom regions can be regarded as "simultaneous" with E.

past of E is bounded by the incoming light lines. The remaining events (the top and bottom regions in Figure A.3) can be regarded as simultaneous with E. More precisely, they are simultaneous for particular choices of world-line. There will be more to be said about this in Section A.3 below. Now light travels very fast indeed and, if common units such as metres and seconds were used, then the light-lines in the diagram would be very close to vertical. To make the diagram comprehensible, units have been used which make the speed of light (usually denoted by the letter c) equal to 1. This puts the light-lines at 45°. For the most part, this book uses these uncommon (aka "natural" or "astronomical") units, with time expressed in terms of years and distance in light-years (the distance travelled by light in one year).

These diagrams both simplify "space" to be 1-dimensional. In fact, of course, it is 3-dimensional and for an accurate diagram it would be necessary to draw it in four dimensions. This is difficult to visualise but you can make a start with a 3-dimensional diagram where space is represented by two dimensions, Figure A.4. In this diagram the outgoing light lines from E fill out a cone called the *light-cone* and the interior of this cone is the *future of E*. Similarly the *past of E* is bounded by the incoming light-cone.

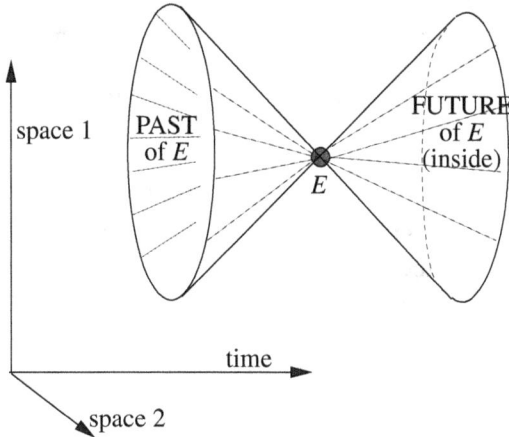

Figure A.4: 3-dimensional causality diagram

There is one other key fact about light that is needed. The speed of light in a vacuum as measured by any observer is always the same. The famous Michelson–Morley experiment was an attempt to find the absolute velocity of the earth through the æther by comparing the speed of light in two perpendicular directions. Later experiments also compared the speeds at two opposite points of the earth's orbit around the sun. In all cases, no difference was found. This negative result could have been explained by assuming that the earth drags the æther with it. The bold explanation which took some time to be accepted was that there is no æther and all observers measure the speed of light to be the same. This bold hypothesis leads to mathematical relativity (aka special relativity) and has been amply justified by the extensive applicability of the theory.

A.3 Lorentz transformations

It is now necessary to explain how the mathematical theory of relativity differs from the simple naive statement that all motion is relative. This is a philosophical truism neatly encapsulated in this anecdote attributed to Wittgenstein:

> *Two philosophers meet in the hall. One says to the other, Why do you suppose people believed for such a long time that the sun goes around the earth, rather than that the earth rotates? The other philosopher replies, Obviously because it looks as though the sun is going around the earth. To which the first philosopher replies, But what would it look like if it looked as though the earth was rotating?*

Motion is always motion measured relative to something else. There is no difference in content between the statement that the sun goes round the earth and the statement that the earth rotates. Both describe the same relative motion. The former is more useful for earth-based purposes whilst the latter is more useful for astronomical purposes. It is not a case that one is *true* and the other *false*. Both are valid. This is the essence of a famous principle, known as "Mach's principle", which is explained in Chapter 4.

But this is not Mathematical Relativity. Mathematical Relativity is a theory which squares the naive principle that all motion is relative with the apparently contradictory fact that the speed of light in a vacuum as measured by any observer is always the same.

The apparent contradiction is because if you measure the speed of a beam of light coming from you to me and if I am moving towards you, then I must measure the same beam travelling more quickly since my speed must be added to the speed of light from you. The resolution of this contradiction is that either my time is different from yours OR my measuring rods are shrinking with respect to yours because of my motion. In Mathematical Relativity BOTH these changes occur. At this point it is necessary to be a bit technical.

Make a simplifying assumption. Suppose that we are in a universe comprising 1 dimension of space and 1 of time and agree to use natural units so that $c = 1$. In other words we are in the universe illustrated in Figure A.3. Suppose that you are at the "origin" at time zero (the point in the middle labelled O in Figure A.5) and not moving . This means that your world-line is the horizontal line to the right in the figure. Suppose for simplicity that I am also at the origin at time zero, but that I am travelling upwards with constant velocity. Then my world-line will be a straight line inclined upwards as illustrated. But as far as I am concerned, it is I who am stationary and you who are travelling (downwards). My view of things is shown in red on the diagram.

Now as far as we are concerned, our notion of "time" corresponds to our motion along our world-lines, so you can think of the lines labelled "world-line" as labelled "time" — the time for the observer moving along that world-line. What does our corresponding "space" look like? For you at rest, space is obviously the vertical line through the origin. The points of this line represent events that are simultaneous with O (from your point of view). But from my point of view space is represented by a line inclined to the right as illustrated in Figure A.6. In other words the events that I see as simultaneous with O are not the same as the events that you see. In Figure A.2, for simplicity, the observers' spaces are drawn as perpendicular to their times. This is true, but it is a peculiar property of the space for Special Relativity, that perpendicular does not always look perpendicular. Figure A.6 is the correct picture.

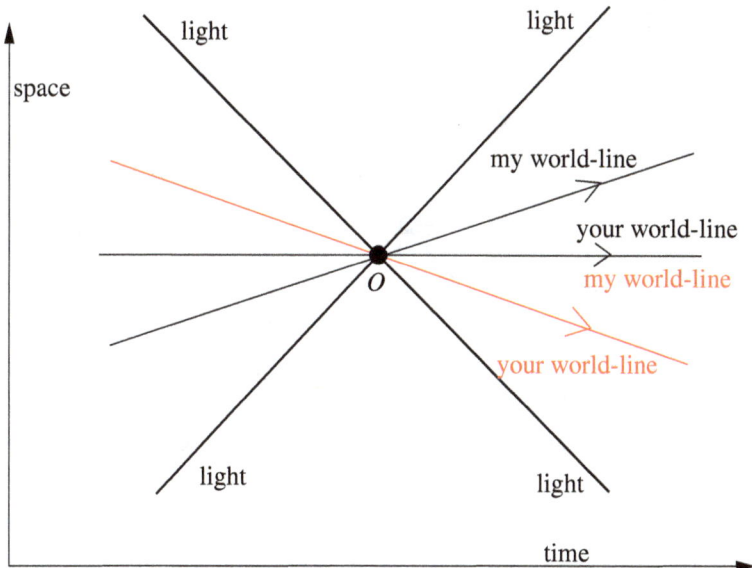

Figure A.5: Our world-lines. My point of view is red, yours is black.

To justify this picture, it is necessary to describe how our two views of the universe are related. From my point of view, I see my space as perpendicular to my time (as you do from your point of view). We can both make an accurate map of the world but a different map. The key to understanding special relativity is to understand how to compare our two maps. On your map a typical event has coordinates (t, x) say but on my map the same event has (usually) different coordinates (t', x'). The transformation that takes (t, x) to (t', x') is indicated roughly in blue in Figure A.6. It takes my time and space axes to yours. The transformation has been drawn as if it was a rotation. Indeed it is, but it is a strange hyperbolic rotation. The fundamental fact that needed is that we agree on light lines. Look now at Figure A.7.

Upward light lines are drawn in green and downward ones in red. They form a grid which covers the whole map. We agree on this grid, but we do not need to agree on the spacing of lines in this grid. We also agree on the

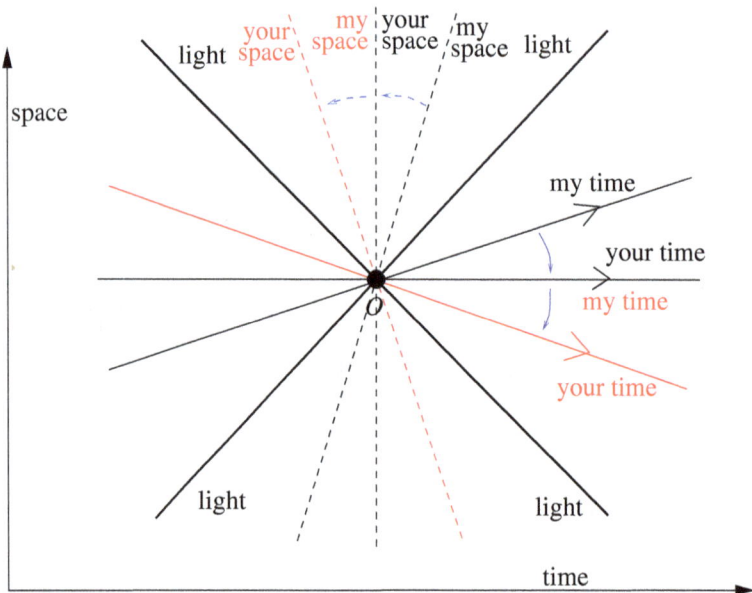

Figure A.6: Our space-times. My point of view is red, yours is black. The transformation taking my view to yours is blue.

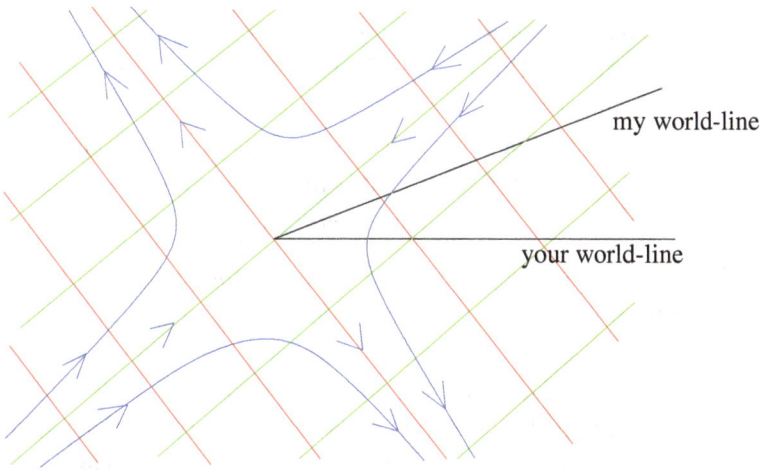

Figure A.7: The light grid and hyperbolic rotation

central point (the event where I meet you). The transformation that takes my view to yours (and takes my world-line to yours) must shrink the green lines and stretch the red ones. We expect the transformation to be uniform over the whole plane (this is justified by thinking of observers moving along parallel paths in space-time) and therefore the stretch of the red lines is by a constant factor k say where $k > 1$ and the shrinkage of the green lines is by another constant $l < 1$. But if we reverse our roles, the transformation that takes your view to mine stretches green lines by $1/l$ and shrinks red ones by $1/k$. But our roles are exactly symmetric and it must be the case that $k = 1/l$.

This shrinkage and stretching has been indicated by blue arrows in the diagram. The formula for this transformation is $(u, w) \mapsto (ku, w/k)$ where (u, w) are the grid coordinates. Since u and w are transformed by reciprocal factors, the transformation preserves the curves given by $uv = C$ for C constant. These are rectangular hyperbolae. In the figure, two of these have been drawn (in blue) corresponding to $C = -1$ (the right and left curved lines) and $C = +1$ (the top and bottom curved lines). In terms of usual coordinates (x for space and t for time) since $u = x + t$ and $v = x - t$ these curves are given by $x^2 - t^2 = C$ for varying C.

The transformation makes points flow along these hyperbolae, as indicated by more blue arrows. This is why this transformation is called a hyperbolic rotation. Now it is evident why my "space" which is transformed to yours by this rotation is inclined upwards to the right (as drawn it in Figure A.6) and therefore any point in the upper or lower quadrants in Figure A.3 are simultaneous with O for a suitable observer as also claimed earlier.

The hyperbolic rotation just arrived at is a simple example of a Lorentz transformation. A general Lorentz transformation is a combination of hyperbolic rotations with ordinary translations and rotations. It is necessary to think of space as 3-dimensional instead of 1-dimensional. This 3-space can be moved around for different points of view by translating and rotating (so called Euclidean motions) and time can also be translated. All these motions together with hyperbolic rotations make up the set of Lorentz transformations (the Lorentz group). Incidentally the Lorentz group has been derived assuming that a Lorentz transformation preserves all light lines and is also uniform. There is a famous theorem of Christopher Zeeman which says that you only need to consider the most basic fact that two observers must agree on, namely causality, to obtain the Lorentz group [106].

A.4 Time dilation and length contraction

Now that there is a good picture of the relationship between our two views of the world, it is possible to explain that other strange property of motion in special relativity, namely that motion causes time to appear to dilate and lengths to appear to contract. Look now at Figure A.8. Suppose I move along my world-line from O to P. From your point of view this takes a time equal to the length OT. But applying the transformation that takes my view to yours, then P moves to an point T' *closer* to O, as drawn. So the real elapsed time *for me* is the length OT' which is smaller than OT. *You see my time dilated.* But of course the situation is symmetric as always, so *I see your time dilated by exactly the same factor.* To see how length contraction works, suppose that my motion happens because I am at the back of a train moving upwards. The front of the train moves on a parallel world line, drawn dashed. When the back of the train is at O in my space (which is the same as the train's space) the front is at F. For you the length of the train is OF'. But applying the transformation taking my space to yours the train really has length OF'' which is larger. *You see my lengths contracted.* Again by symmetry, *I see your lengths contracted.*

Precise formulae for these dilation/contraction effects can be derived from Figure A.8. The hyperbola has formula $t^2 - x^2 = C$ for some C and, since T' lies on it, $C = u^2$ so $t^2 - x^2 = u^2$. Dividing by t^2 gives $1 - (x/t)^2 = (u/t)^2$ or $1 - v^2 = (1/d)^2$ where $v = x/t$ is our mutual velocity and $d = t/u$ is the dilation factor for time. Thus time is dilated by $d = 1/\sqrt{1 - v^2}$ and dually lengths are contracted by $1/d = \sqrt{1 - v^2}$. These are the formulae in natural units (with $c = 1$). In common units the formulae are: time dilation $d = 1/\sqrt{1 - v^2/c^2}$ and length contraction $1/d = \sqrt{1 - v^2/c^2}$.

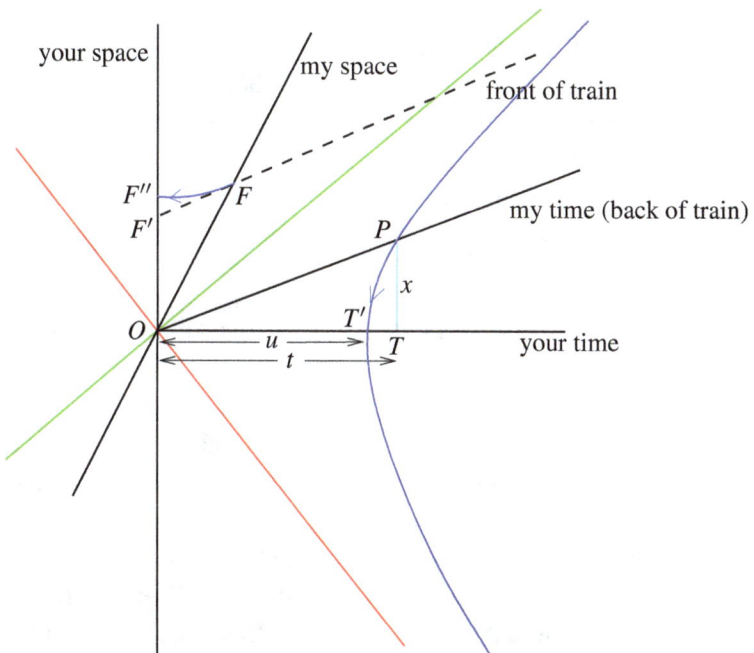

Figure A.8: Time dilation and length contraction

A.5   Minkowski space

At this point a good description of a particular space-time which is called Minkowski space has been achieved. This space is the fundamental space-time for relativity. Special relativity takes place in Minkowski space and general relativity is built upon it as will be seen in the next section.

It is necessary to be very precise. Minkowski space is 4-dimensional space with coordinates (t, x, y, z) where the first coordinate t is called *time* and the other three are *space*. The transformations just called Lorentz transformations act on Minkowski space and special relativity is the study of properties which are unchanged by Lorentz transformations. Home in on one particular property, namely *length*. As usual, for simplicity, assume there is just one dimension of space x. When the study of length in 2-dimensional Minkowski space is finished, it will be easy to generalise back to 4 dimensions.

Consider two events (points of space-time) in Minkowski space with coordinates (t, x) and (t', x'). The fundamental "property" of the two events taken together is the "number" s where $s^2 = -(t - t')^2 + (x - x')^2$. This has been put in inverted commas because sometimes s is the square root of a negative number, in other words it may be imaginary. To avoid having to think about imaginary numbers use s^2 instead of s. s^2 is the appropriate number to be considered "length" (or rather the square of length) in Minkowski space.

First notice that it doesn't change under translation — ie replacing x by $x - a$ and t by $t - b$ where a and b are constants. Translate so that one of the points (say (t', x')) is at the origin $(0, 0)$. Then the formula for s^2 is simpler $s^2 = -t^2 + x^2$. Now consider a hyperbolic rotation. As found above, this preserves rectangular hyperbolae $x^2 - t^2 = C$. In other words it preserves s^2 (and hence s). So s^2 is preserved by all Lorentz transformations as "length" must. There is an obvious analogy with Cartesian (ordinary) length s in Euclidean (normal) space, which by Pythagoras' Theorem has the formula $s^2 = x^2 + y^2$. But there are obvious differences — it can be zero, for example if $x = \pm t$, ie if (t, x) lies on a light line through the origin. Lengths in ordinary space are never zero!

Consider some other cases. Suppose that t is positive and that $t > x$ or $t > -x$ in other words that (t, x) lies in the right hand quadrant. Then $s^2 = -t^2 + x^2$ is negative (length is imaginary if you like). A similar thing happens if (t, x) lies in the left hand quadrant. If $t < x$ or $-t > x$ (top quadrant) s^2 is positive. Finally if $x = t$ or $x = -t$ (the light lines through the origin) then s^2 is zero. These facts are illustrated in Figure A.9.

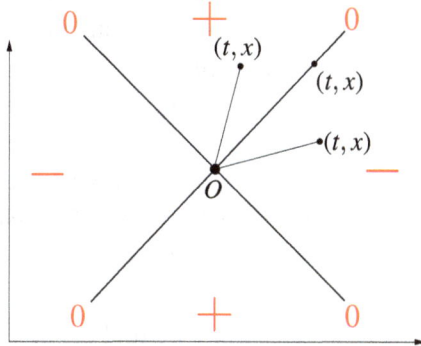

Figure A.9: The sign of s^2

Causality can be interpreted in terms of this new concept of "length". Two events P and Q are causally related if the (square) of the length of the interval PQ is zero or negative and this length can be thought of as being the time separating the two events. Similarly if the (square) of the length of the interval PQ is positive then the events are not causally related and we can think of the length as being the distance in space separating the two events.

This length is the *metric* on Minkowski space. More precisely, moving back now to 4 dimensions, Minkowski space is 4-dimensional space with coordinates (t, x, y, z) and metric (distance) s given by

$$s^2 = -t^2 + x^2 + y^2 + z^2.$$

Metrics are often expressed in infinitesimal form using ds (a tiny step along s) etc:

$$ds^2 = -dt^2 + dx^2 + dy^2 + dz^2$$

A.6   General Relativity

Now move on to consider mathematical models for the universe. Minkowski space is the simplest model but it is far too simple. Einstein's deep insight was that the force of gravity — the force that keeps us anchored to the earth and which keeps the earth moving around the sun — should be thought of as encoded in the fabric of space-time by means of curvature. With this insight, Minkowski space (which has no curvature) is a model for an empty universe: one with no planets, stars or galaxies. So more general models are needed. Nevertheless Minkowski space remains fundamentally important because it correctly describes the local geometry of the universe. It is accurate over small distances and for a small interval of time. This fact is taken as an axiom in general relativity. What it says in words is that the small scale geometry of space-time is the same everywhere for all observers. And of course, these local Minkowski spaces are inertial frames.

A.7   Manifolds and space-times

A *manifold* is a space which is locally the same as ordinary (Euclidean) space, but which might be quite different globally. The dimension is the dimension of the local Euclidean space. A one dimensional maniold is locally like a line but could be closed (as a circle). A 2-manifold is a surface of which the sphere and the torus are examples (Figure A.10).

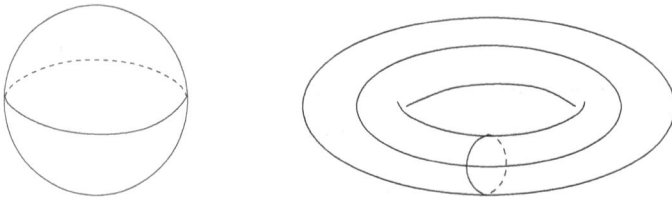

Figure A.10: Sphere (left) and torus (right)

Here is a simple example of a 3-manifold which is not ordinary 3-space. Think of a cube in 3-space and make a conceptual leap by assuming that the top of the cube is exactly the same as the bottom. What this means is that if you move upwards through the top, you immediately appear at the bottom. Make the same leap for the other two pairs of oposite faces, so that if you move out through the left side you immediately appear at the right and similarly forwards and backwards. The space being described is the 3-manifold known as the 3-*torus*.

4-dimensionsal manifolds are needed for relativity. But with the essential difference that at any point some directions are time-like and some space-like. A *space-time*, also called a *Lorentz* manifold, is a space locally like Minkowski space. At each point (event) there are two null cones representing incoming and outgoing light lines as in figure A.4. The 3-torus can be made into a space-time by adding one extra dimension for time and using Minkowski space as a model for how this time dimension fits with the three space dimensions.

There is one other important property of the manifolds used for relativity. They are *smooth* manifolds, which means they have a smooth metric which is locally diffeomorphic (smoothly equivalent) to the metric on Minkowski space. Thus the light-cones at each point can be defined, as in Minkowski space, as directions in which the metric is null (points on the same light ray have zero separation in the metric). Further time-like directions are ones where s^2 (the square of the metric) is negative and space-like ones where it is positive.

The notation used for a general metric is

$$ds^2 = \sum_{i,j} g_{i,j}\, dx_i\, dx_j$$

and the array of coefficients $\mathbf{g} = (g_{i,j})$ is also called the metric. In technical terms, $\mathbf{g}$ is a bilinear form of index $(-1, 3)$. Here -1 is for the time-like direction and 3 for the three space-like directions. An important example is the Schwarzschild metric which is used repeatedly through the book

(A.1) $ds^2 = -(1-2M/r)\, dt^2 + (1/(1-2M/r))\, dr^2 + r^2\, (d\theta^2 + \sin^2\theta\, d\phi^2).$

Here M is a constant interpreted as central mass and spherical coordinates (r, θ, ϕ) are used for space. In this example $g_{tt} = -(1 - 2M/r)$, $g_{rr} = 1/(-g_{tt})$, $g_{\theta\theta} = r^2$, $g_{\phi\phi} = r^2 \sin^2\theta$ and the others are zero. The manifold for this metric is ordinary 3-space with the origin removed crossed with one dimension (for time).

A.8   Curvature

To proceed it is necessary to discuss the *curvature* of a space-time. This idea applies to any manifold and the simplest example to think about is the curvature of a *surface* (or 2-manifold). The most familiar curved surface is the *sphere* or the surface of a round ball. It is obviously curved, but to explain curvature in general it is necessary to encapsulate curvature in mathematical terms. Think about a triangle in the sphere and think about carrying a vector around that triangle keeping it as parallel to itself as possible (this is called *parallel transport*). Whatever triangle you choose, the vector ends up pointing in a different direction. For example see Figure A.11.

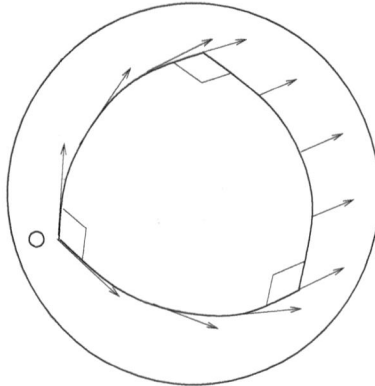

Figure A.11: Transporting a vector around a trangle on the sphere (start at O)

Riemann curvature

It is not necessary to use a triangle to detect the curvature. Transporting a vector around almost any closed curve on the sphere results in a non-parallel vector. The same idea can be used in any manifold with a metric. By transporting vectors around small curves it is possible to define curvature. If the curve is chosen to lie in a plane, this gives the notion of the curvature of that plane. But notice that this is a vector — the discrepancy after transport — not a number.

The *Riemann curvature* of a manifold is this idea used exhaustively. A space-time is a 4-manifold and, sticking to coordinate planes gives a choice of 12 planes. In each plane a coordinate vector can be transported around a small curve and the discrepancy, which is a vector, read. The Riemann curvature tensor, R^i_{jkl}, is an array of 4^4 numbers obtained in this way. The definition of R^i_{jkl} is: transport the j th coordinate vector around a small curve in the (k, l)-plane and read the change in the i th coordinate of the result. It is a $(1, 3)$-*tensor* because of the way it transforms under change of coordinates. There are many symmetries and identities amongst the components and there are in fact only 20 independent components. The Riemann curvature gives all possible information about how the manifold curves. However for Einstein's General Relativity, only about half of this information is needed, namely the Ricci curvature.

There are (fairly complicated) formulae for the Riemann curvature in terms of the metric:

(A.2)
$$R^i_{jkl} = \partial_k \Gamma^i_{lj} - \partial_l \Gamma^i_{kl} + \sum_\lambda (\Gamma^i_{k\lambda} \Gamma^\lambda_{lj} - \Gamma^i_{l\lambda} \Gamma^\lambda_{kj})$$

where the *Christoffel symbols* Γ^m_{ij} are defined by

(A.3)
$$\Gamma^m_{ij} = \sum_k \tfrac{1}{2} g^{km} (\partial_i g_{kj} + \partial_j g_{ik} - \partial_k g_{ij}),$$

$\partial_i = \partial/\partial x_i$ means differentiation wrt to x_j and g^{ij} is the inverse matrix to g_{ij}. These formulae are useful in special cases (eg for diagonal metrics where many terms vanish).

Ricci curvature

The *Ricci curvature* is a contraction of the Riemann curvature. It is another tensor (in fact a 2-tensor or bilinear form) and the definition is $\mathrm{Ric}_{ij} = \sum_k R^k_{ikj}$. There are again symmetries and the number of independent components is 12 rather than 16. It has a simple geometric interpretation.

A bilinear form is determined by values on single repeated vectors (rather than general pairs of vectors) — the associated quadratic form — and from the definition this has the following meaning. Consider 2-planes containing the given vector v and add the curvature for 4 mutually perpendicular planes. So this is the "average" curvature for planes containing v. But there is a simpler interpretation. Consider a small cone of vectors near to v and measure the 4-volume of this cone. It differs from the result in flat (Minkowski) space due to curvature. This difference is the value of (the associated quadratic form to) Ric on v. So Ricci curvature measures the way space-time expands (or contracts).

The diagonal components of the Ricci curvature are the sectional curvatures which are directly analogous to the curvature of a surface; these are the curvatures of four mutually perpendicular hyperplanes measured in a perpendicular direction. The contraction of Ric is the *scalar curvature S* defined by

$$S = \sum_{i,j} g^{ij} \operatorname{Ric}_{ij}.$$

A.9   Einstein's equations

Einstein's idea of pure genius was to interpret the force of gravity as due to curvature of space-time. As has been seen, the Ricci curvature determines the way volume grows. If this is positive then nearby parallel geodesics will tend to converge (as if under the influence of a force). The formulation that Einstein eventually found after much effort was in terms of this, the Ricci curvature, rather than the general Riemann curvature. The *Einstein tensor* denoted G_{ij} is not quite the Ricci curvature. Einstein's equations express the curvature in terms of the presence of matter. There is a *stess-energy* tensor T which encodes the energy and momentum of matter. The idea was that the equations should say that $G = kT$ for some suitable constant k. Conservation of energy and momentum implies that div $T = 0$ where div is divergence. But div Ric is non zero, in fact it is $\frac{1}{2}dS$, half the deriative of the scalar curvature, so to achieve div $G = 0$, define the Einstein tensor G to be $\operatorname{Ric} - \frac{1}{2}S\mathbf{g}$, ie $G_{ij} = \operatorname{Ric}_{ij} - \frac{1}{2}Sg_{ij}$ where S is scalar curvature and $\mathbf{g}$ is the metric.

Einstein's equations now read:

$$G = 8\pi T$$

The constant 8π is found by considering simple special cases and using natural units where Newton's gravitational constant (also confusingly denoted G) is 1. T will not be described explicitly here because, for the most part, this book is concerned with vacuum solutions ($T = 0$) or modifications of these due to inertial effects. The interested reader can find many good descriptions in the literature. The vacuum equations are $G = \text{Ric} - \frac{1}{2}S\mathbf{g} = 0$. But contracting this equation implies that $S = 0$ and hence:

Einstein's vacuum equations are equivalent to $\text{Ric} = 0$.

Einstein's biggest blunder

In order to have a static solution for the universe, Einstein modified his basic equations by adding a *cosmological constant* κ times $\mathbf{g}$ to his tensor:

$$G + \kappa\mathbf{g} = 8\pi T$$

or

(A.4) $$\text{Ric} + (\kappa - \tfrac{1}{2}S)\mathbf{g} = 8\pi T.$$

This happened before the observations of Slipher and Hubble–Humason suggested that the universe might not be static but expanding. Einstein then rescinded his cosmological constant κ calling this his biggest blunder. If he hadn't introduced it, he could have predicted the observed expansion! Since the 1998 WMAP observations, most cosmologists are happy to keep the cosmological constant since the universe seems now to approximate de Sitter space which has a positive cosmological constant (as will be seen shortly). From the author's point of view, Einstein's biggest blunder was the reintroduction of a universal time in his (and consequently current mainstream) models for the universe in the large. There is no universal time in either special or general relativity. It is the assumption of a universal time that leads to the (false) Big Bang theory which dominates current cosmology.

Vacuum equations with cosmological constant

For the case of a vacuum ($T = 0$) the **g** terms in equation A.4 can be collected to give

(A.5) $\text{Ric} = \Lambda \mathbf{g}$

where $\Lambda = \frac{1}{2}S - \kappa$ is a scalar field. This formulation is slightly more general than Einstein's since it allows κ to vary over space-time.

The Schwarzschild and de Sitter solutions

Finding general solutions to the Einstein equations is not easy because of their complication when expressed in terms of the metric, but there is an important special case when it is fairly easy. This is the spherically-symmetric case and is the appropriate case for studying the metric near an isolated heavy body. Spherical symmetry implies that the metric can be expressed the in the form:

(A.6) $ds^2 = -Q\,dt^2 + P\,dr^2 + r^2\,d\Omega^2$

where P and Q are positive functions of r and t on a suitable domain. Here t is time, r is "distance from the centre" and $d\Omega^2$, the standard metric on the unit 2-sphere S^2, is an abbreviation for $d\theta^2 + \sin^2\theta\,d\phi^2$. This metric is diagonal which implies that many of the terms in Equations A.2 and A.3 are zero and it is not too hard to compute the Ricci curvature, see for example Win [104]. Then it is fairly easy to prove that if Equation A.5 holds then P and Q are independent of t, Λ is constant and

$$Q = \frac{1}{P} = 1 - \frac{\Lambda r^2}{3} - \frac{2M}{r}$$

with M constant. For details here see [87]. This is mild generalisation of Birkhoff's theorem.

The special case $\Lambda = 0$ is the *Schwarzschild metric* and the case $M = 0$ is the *de Sitter metric*. The general case is the Schwarzschild–de Sitter metric also called the Kottler metric.

Black holes

The Schwarzschild metric is the unique spherically-symmetric metric satisfying Einstein's vacuum equations without a cosmological constant. It is given by A.6 with $Q = 1/P = 1 - 2M/r$. The metric appears to go singular at $r = 2M$ (the *Schwarzschild radius*) where $P = 1/(1 - 2M/r)$ is infinite. The solution was discovered in 1915 just a few months after Einstein published his theory and for nine years it was believed that this singularity was a real property of the space and the boundary $r = 2M$ separated real space (outside the Schwarzschild radius) from the virtual space inside. This belief continued until in 1924, when Arthur Eddington showed that the singularity disappeared after a suitable change of coordinates. Nevertheless the Schwarzschild boundary has a real significance for a distant observer. A photon starting at or inside the Schwarzschild boundary cannot cross this boundary. The whole of the future of an event on the boundary lies inside the Schwarzschild radius. To a stationary outside observer the boundary appears completely black — a *black hole* in fact.

Black holes have captured the imagination of the general scientific public and many good treatments of them can be found in the literature to follow up the bare bones given here.

De Sitter space

The de Sitter metric defines a space called de Sitter space. It is of fundamental importance for the geometry of the universe described in this book because the new model for the universe with observed redshift is based on it (see Section 9.4) and also the new explanations for the CMB and gamma ray bursts (see Section 9.5 and Section 9.6).

This space is explored in some detail in Section 9.3 as part of the explanation of redshift. A fuller treatment can be found in Appendix B.

Appendix B

De Sitter space

B.1 Minkowski space

Minkowski n-space M^n is $\mathbb{R}^n = \mathbb{R} \times \mathbb{R}^{n-1}$ (time times space) equipped with the standard (pseudo)-metric of signature $(-, +, \ldots, +)$:

$$ds^2 = -dx_0^2 + dx_1^2 + \ldots + dx_{n-1}^2$$

The time coordinate is x_0 and the space coordinates are $x_1, \ldots, x_{n-1}$. For $x, y \in M^n$, the (pseudo-)inner product $\langle x, y \rangle$ is defined to be $-x_0 y_0 + x_1 y_1 + \ldots + x_{n-1} y_{n-1}$.

The *Lorentz n-group* is the group of "isometries" (transformations preserving the inner product) of Minkowski space, fixing $\mathbf{0}$, and preserving the time direction. This implies that a Lorentz transformation is an linear isomorphism of $\mathbb{R}^n$ as a vector space. If, in addition to preserving the time direction, it also preserves space orientation then the resulting group can be denoted $SO(1, n-1)$. Notice that a Lorentz transformation which preserves the x_0-axis is an othogonal transformation of the perpendicular $(n-1)$-space, thus $SO(n-1)$ is a subgroup of $SO(1, n-1)$ and elements of this subgroup are (Euclidean) rotations about the x_0-axis.

Minkowski 4-space is simply called *Minkowski space* and is the simplest example of a space-time. The Lorentz 4-group is called the *Lorentz group*.

B.2   Space-times

A pseudo-Riemannian manifold L is a manifold equipped with non-degenerate quadratic form g on its tangent bundle called the *metric*. A *space-time* is a pseudo-Riemannian 4-manifold equipped with a metric of signature $(-,+,+,+)$. Minkowski space is the simplest example of a space-time and in general the Lorentz group acts as structure group for the tangent bundle of a space-time. The metric is often written as ds^2, a symmetric quadratic expression in differential 1-forms as above. A tangent vector v is *time-like* if $g(v) < 0$, *space-like* if $g(v) > 0$ and *null* if $g(v) = 0$. The set of null vectors at a point form the *light-cone* at that point and this is a cone on two copies of S^2. The set of time-like vectors at a point breaks into two components bounded by the two components of the light cone. A choice of one of these components determines the *future* at that point and *time orientability*, ie a global choice of future pointing light cones is always assumed. An *observer field* on a space-time L is a smooth future-oriented time-like unit vector field on L.

B.3   de Sitter and hyperbolic spaces

Now go up one dimension. *Hyperbolic 4-space* is the subset
$$\mathbb{H}^4 = \{\langle x, x \rangle = -a^2, \; x_0 > 0 \mid x \in M^5\}.$$
de Sitter space is the subset
$$\mathrm{deS} = \{\langle x, x \rangle = a^2 \mid x \in M^5\}.$$
There is an isometric copy $\mathbb{H}^4_q$ of hyperbolic space with $x_0 < 0$. The induced metric on hyperbolic space is Riemannian and on de Sitter space is Lorentzian. Thus de Sitter space is a space-time. It is a solution of Einstein's equations with positive cosmological constant $\Lambda = 3/a^2$ and no matter.

The *light cone* is the subset
$$L = \{\langle x, x \rangle = 0 \mid x \in M^5\}.$$
and is the cone on two 3-spheres with natural conformal geometries (see hyperbolic geometry below). These are S^3 and S^3_q where S^3 is in the positive time direction and S^3_q negative.

B.4 Projective geometry

Points of

(B.1) $$S^3 \cup S_q^3 \cup \mathrm{deS} \cup \mathbb{H}^4 \cup \mathbb{H}_q^4$$

are in natural bijection with half-rays from the origin and this is called *half-ray space*. Considering full rays (lines) through the origin gives a copy of projective 4-space $\mathbb{P}^4$ which has half-ray space as its unique double cover. $SO(1,4)$ acts faithfully on $\mathbb{P}^4$ by projective transformations. Each of $S^3, S_q^3, \mathbb{H}^4$ and $\mathbb{H}_q^4$ is faithfully represented as a subset of $\mathbb{P}^4$ with copies identified. $SO(1,4)$ acts on $\mathbb{H}^4$ by hyperbolic isometries (see below). The natural linear structure on $\mathbb{P}^4$ (given by subspaces of $\mathbb{R}^5$) induces a natural linear structure on each of $\mathbb{H}^4$ and deS and in particular planes through the origin cut $\mathbb{H}^4$ and deS in geodesics and all geodesics are of this form (this is proved below).

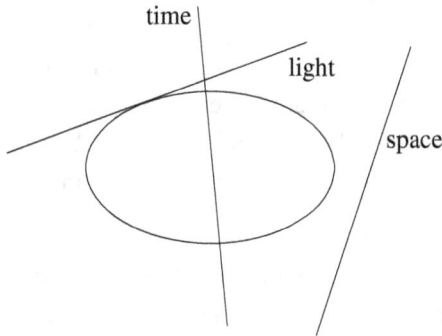

Figure B.1: Three types of lines

Geodesics come in three types: *light-like* corresponding to planes tangent to the light cone; *time-like* corresponding to planes which intersect $\mathbb{H}^4$ and *space-like* corresponding to planes which meet L only at the origin (and hence miss $\mathbb{H}^4$). Figure B.1 is a projective picture illustrating these types.

B.5   Hyperbolic and de Sitter geometry

The subset $\mathbb{H}^4$ of $\mathbb{P}^4$ with the action of $SO(1,4)$ is the *Klein model* of
hyperbolic 4-space. S^3 is then the sphere at infinity and $SO(1,4)$ acts by
conformal transformations of S^3 and indeed is isomorphic to the group
of such transformations. $SO(1,4)$ acts as the group of time and space
orientation preserving isometries of deS and is also known as the *de Sitter
group* as a result.

B.6   Transitivity of points and geodesics

There is an element of $SO(1,4)$ carrying any half-ray to any other of the
same type (corresponding to the decomposition (B.1)). Here is an explicit
way to see this which also proves transitivity on geodesics and checks the
characterisation of geodesics mentioned earlier. A Lorentz transformation
of M^2, called a *shear*, namely $x \mapsto \left(\begin{smallmatrix} \cosh q & \sinh q \\ \sinh q & \cosh q \end{smallmatrix} \right) x$, for suitable choice of
$q \in \mathbb{R}$ acting in a vertical plane (one containing the x_0-axis) and crossed
with the identity on the perpendicular 3-space, will move any point of $\mathbb{H}^4$
to the centre (intersection with the x_0-axis) and any point of deS to a point
on the equator (intersection with the (x_1, x_2, x_3, x_4)-space). Then a rotation
about the x_0-axis carries it to any other point. This proves transitivity for
points in $\mathbb{H}^4$ (and similarly $\mathbb{H}_q^4$) and deS. For S^3 (and similarly S_q^3) a
rotation about the x_0-axis carries one point to any other.

This argument is now extended to prove transitivity on geodesics. Since it
has not yet proved that geodesics are intersections with planes through the
origin, such intersections are called *lines* and it will be seen that lines are in
fact geodesics shortly. Notice that a time-like plane (one meeting $\mathbb{H}^4$, see
terminology introduced above) meets deS in *two* antipodally opposite lines
which get identified in $\mathbb{P}^4$ and which we call a line-pair.

Transitivity will be proved for lines of the same type. In $\mathbb{H}^4$ all lines are
time-like and any point of each line can be moved to the centre and then
a rotation carries one line into the other. Turning now to de Sitter space,
the same sequence of transformations applied to deS carries any time-like

line-pair to any other and a rotation can be used to swap the lines if necessary. A space-like line can be carried to the equator by two perpendicular shears.

To prove transitivity for light-like lines, observe that in Euclidean terms deS is a hyperboloid of one sheet ruled by lines and that each tangent plane to the light cone meets deS in two ruling lines. These lines are light-lines in M^5 and hence in deS. (See Figure B.2, which is taken from Moschella [77].) Now a point on a given light-line can be carried to a point of the equator by a shear followed by a rotation. But the rotations fixing this point now carry the light-line around the light cone in deS and hence any two are equivalent under an isometry.

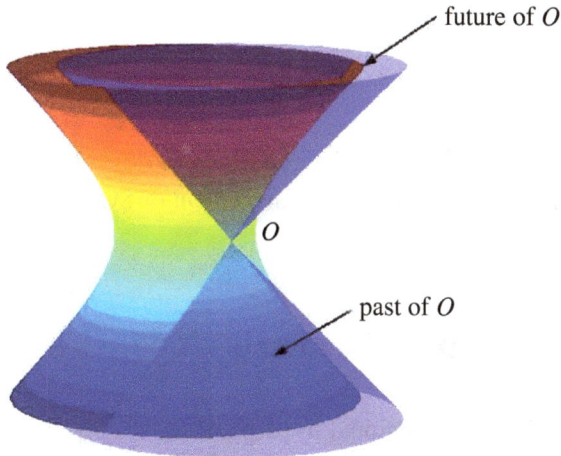

Figure B.2: Light-cones: the forward light-cone in deS is the cone on a 0-sphere (two points) in the dimension illustrated (ditto backwards), in fact each is the cone on a 2-sphere. The figure is reproduced from [77].

The proof of transitivity shows that there are isometries which move any line along itself and further it can now be seen that the isometries which fix a given line point-wise form a SO(3) subgroup of rotations. Thus by symmetry, parallel transport along a line carries a line into itself and they are, as claimed, geodesics. Since there are lines through each point in each direction, all geodesics are of this form. Thus:

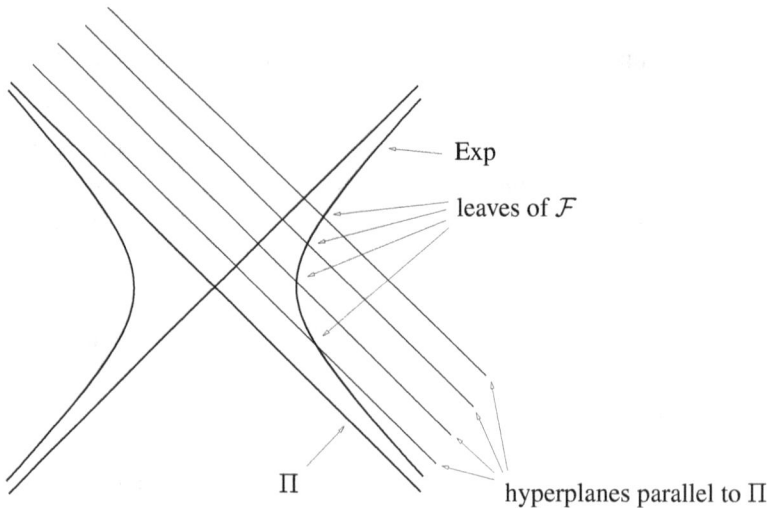

Figure B.3: The foliation $\mathcal{F}$ in the (x_0, x_4)-plane

Proposition B.1 *Let l, m be geodesics in deS of the same type and let $P \in l$ and $Q \in m$. Then there is an element of $SO(1, 4)$ carrying l to m and P to Q.*

B.7   The expansive metric

Let Π be the 4-dimensional hyperplane $x_0 + x_4 = 0$. This cuts deS into two identical regions. Concentrate on the upper complemetary region Exp defined by $x_0 + x_4 > 0$. Π is tangent to both spheres at infinity S^3 and S_q^3. Name the points of tangency as P on S^3 and P' on S_q^3. The hyperplanes parallel to Π, given by $x_0 + x_4 = k$ for $k > 0$, are also all tangent to S^3 and S_q^3 at P, P' and foliate Exp by paraboloids. Denote this foliation by $\mathcal{F}$. It will be seen that each leaf of $\mathcal{F}$ is in fact isometric to $\mathbb{R}^3$. There is a transverse foliation by the time-like geodesics passing through P and P'.

These foliations are illustrated in Figures B.3 and B.4. Figure B.3 is the slice by the (x_0, x_4)-coordinate plane and Figure B.4 (the left-hand figure) shows the view from the x_4-axis in 3-dimensional Minkowski space (2-dimensional de Sitter space). This figure and its companion are again taken from Moschella [77].

Let G be the subgroup of the Lorentz group which fixes P (and hence P' and Π). G acts on Exp. It preserves both foliations: for the second foliation this is obvious, but all Lorentz transformations are affine and hance carry parallel hyperplanes to parallel hyperplanes; this proves that it preserves the first foliation. Furthermore affine considerations also imply that it acts on the leaves of $\mathcal{F}$ by scaling from the origin: in other words there is a scale factor $\lambda(g)$ for each $g \in G$ which maps the leaf at distance μ from Π to the one at distance $\lambda(g)\mu$ (here distance means Euclidean distance in $\mathbb{R}^5$). The map $\lambda \colon G \to \mathbb{R}_*$ is a homomorphism from G to the positive reals under multiplication.

Now consider the action of G on S^3 (the light sphere at infinity). As remarked earlier $SO(1,4)$ acts on S^3 as the group of conformal isomorphisms. Thus G is the subgroup of conformal isomorphisms of S^3 fixing the point P. But this is a very familiar group. Use stereographic projection to identify $S^3 - P$ with $\mathbb{R}^3$ then G acts as the group of conformal isomorphisms of $\mathbb{R}^3$, which is precisely the group of similarities of $\mathbb{R}^3$ ie the group generated by isometries and dilations. There is then another homomorphism $\sigma \colon G \to \mathbb{R}_*$ which maps each element to its scale factor. It can be checked that $\lambda = \sigma$. One way to see this is to list the non-trivial normal subgroups of G (this is not hard) and check that there is only one homomorphism to $\mathbb{R}_*$. Another

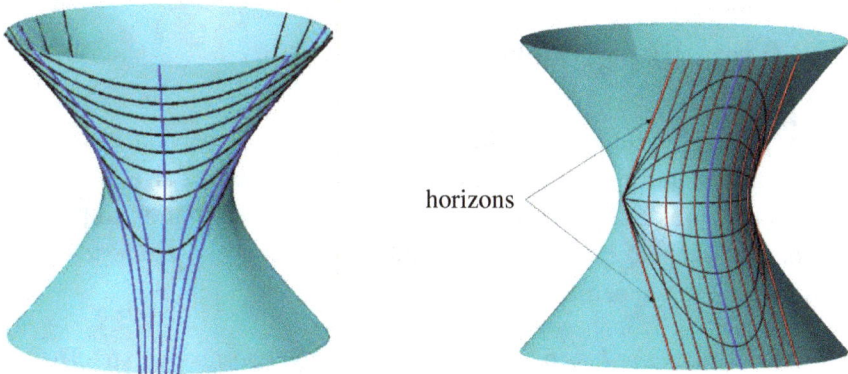

Figure B.4: Two figures reproduced from [77]. The left-hand figure shows the foliation $\mathcal{F}$ (black lines) and the transverse foliation by geodesics (blue lines). The righthand figure shows the de Sitter metric as a subset of deS.

way is to check directly that a standard shear j, say, in the (x_0, x_4)-plane is a pure dilation at infinity by exactly $\lambda(j)$. (It is easy to see that j acts on leaves by dilation to first order and at infinity. The scaling is given by affine considerations.)

Now define H to be the kernel of $\lambda = \sigma$. This comprises elements of G which maps the leaves of $\mathcal{F}$ into themselves, and in terms of the action on $S^3 - P = \mathbb{R}^3$, it is the subgroup of Euclidean isometries. Thus the Euclidean group acts on each leaf of $\mathcal{F}$. By dimension considerations it acts an the full group of isometries of each leaf. It follows that each leaf has a flat Euclidean metric.

It can now be seen that the metric on Exp is the same as the Robertson–Walker metric for a uniformly expanding infinite universe. The transverse foliation by time-like geodesics determines the standard observer field and the distance beetween hyperplanes defining $\mathcal{F}$ gives a logarithmic measure of time. Coordinates are given below. Notice that it has been proved that every isometry of Exp is induced by an isometry of deS.

It is worth remarking that exactly the same analysis can be carried out for $\mathbb{H}^4$ where the leaves of the foliation given by the same set of hyperplanes are again Euclidean. This gives the usual "half-space" model for hyperbolic geometry with Euclidean horizontal sections and vertical dilation.

B.8   Time-like geodesics in Exp

There is a family of time-like geodesics built in to Exp namely the observer field mentioned above. These geodesics are all *static*. They are all equivalent by a symmetry of Exp because we can use a Euclidean motion to move any point of one leaf into any other point. Other time-like geodesics are *non-static*. Here is a perhaps surprising fact:

Proposition B.2 *Let l, m be any two non-static geodesics in Exp. Then there is an element of $G = \text{Isom}(\text{Exp})$ carrying l to m.*

Thus there in no concept of conserved velocity of a geodesic with respect to the standard observer field in the expansive metric. This fact is important for the analysis of black holes in a universe with a fixed cosmological constant, cf [71].

The proof is easy if one thinks in terms of hyperbolic geometry. Time-like geodesics in Exp are in bijection with geodesics in $\mathbb{H}^4$ since both correspond to 2-planes through the origin which meet $\mathbb{H}^4$. But if using the upper half-space picture for $\mathbb{H}^4$ static means vertical and two non-static geodesics are represented by semi-circles perpendicular to the boundary. Then there is a conformal map of this boundary (ie a similarity transformation) carrying any two points to any two other: translate to make one point coincide and then dilate and rotate to get the other ones to coincide.

B.9 The de Sitter metric

There is another standard metric inside de Sitter space which is the metric which de Sitter himself used. This is illustrated in Figure B.4 on the right. This metric is static, in other words there is a time-like Killing vector field (one whose associated flow is an isometry). The region where it is defined is the intersection of $x_0 + x_4 > 0$ defining Exp with $x_0 - x_4 < 0$ (defining the reflection of Exp in the (x_1, x_2, x_3, x_4)-coordinate hyperplane). The observer field, given by the Killing vector field, has exactly one geodesic leaf, namely the central (blue) geodesic. The other leaves (red) are intersections with parallel planes not passing through the origin. There are two families of symmetries of this subset: an SO(3)-family of rotations about the central geodesic and shear along this geodesic (in the (x_0, x_1)-plane). Both are induced by isometries of deS.

This metric accurately describes the middle distance neighbourhood of a black hole in empty space with a nonzero cosmological constant. The embedding in deS is determined by the choice of central time-like geodesic. Proposition B.2 then implies that there are precisely two types of black hole in a standard uniformly expanding universe.

B.10   Explicit formulae

Here are explicit formulae for these two metrics in terms of the x_i adapted from wikipedia.

Expansive metric   Let

$$x_0 = a \sinh(t/a) + r^2 \exp(t/a)/2a$$
$$x_4 = a \cosh(t/a) - r^2 \exp(t/a)/2a$$
$$x_i = \exp(t/a) y_i \text{ for } 1 \le i \le 3$$

where $r^2 = \sum_i y_i^2$. Then in the (t, y_i) coordinates the metric reads

$$ds^2 = -dt^2 + \exp(2t/a) du^2$$

where $du^2 = \sum dy_i^2$ is the flat metric on y_i's.

de Sitter coordinates (static coordinates)   Let

$$x_0 = \sqrt{a^2 - r^2} \sinh(t/a)$$
$$x_4 = \sqrt{a^2 - r^2} \cosh(t/a)$$
$$x_i = r z_i \text{ for } 1 \le i \le 3$$

where z_i gives the standard embedding of S^2 in $\mathbb{R}^3$ then in r, z_i coordinates the metric reads

(B.2) $$ds^2 = -Q dt^2 + 1/Q dr^2 + r^2 d\Omega^2$$

where $Q = 1 - (r/a)^2$ and $d\Omega^2$ is the standard metric on S^2.

Appendix C

Quasars: technical material

This appendix contains the technical material from the three author paper [88] deferred from Chapter 6. This paper, which is joint work with Robert MacKay and Rosemberg Toala Enriques, is still in draft form: the next version will deal accurately with ionisation of the incoming gas/plasma stream and model the settling region. Note that, throughout this appendix, scientific (MKS) units are used rather than the natural units used elsewhere and that G is Newton's gravitational constant and not the Einstein tensor. The appendix starts with the omitted details for the Bondi sphere radius and the accretion rate.

C.1 Bondi sphere radius and accretion rate

The Bondi sphere of radius B is defined by equating the root mean square velocity $\sqrt{3kT/m_H}$ of Hydrogen atoms in the medium with the escape velocity $\sqrt{2GM/B}$. Here T is the temperature of the medium at the Bondi radius, M is the mass of the BH, G is the gravitational constant, k is Boltzmann's constant and m_H is the mass of a proton. Thus:

$$(C.1) \qquad \qquad B = \frac{2GMm_H}{3kT}$$

Note that the Newtonian formula for escape velocity has been used, which, as will be seen later, is also correct in Schwarzschild geometry.

The significance of the Bondi sphere is that protons in the medium are trapped (on average) inside this sphere because they have KE too small to escape the gravitational field of the BH. The mass of matter per unit time trapped in this way is called the *accretion rate* A and can be calculated as

$$(C.2) \qquad\qquad A = 2B^2 n \sqrt{2\pi kTm_H}$$

where n is the density of the medium (number of protons per unit volume).

Here are the details for this calculation. Maxwell's distribution for the radial velocity v_r has density $\sqrt{m_H/2\pi kT}\, e^{-m_H v^2/2kT}$, so the mean $\bar{v}_r$ over inward velocities is

$$\int_0^\infty 2\sqrt{\frac{m_H}{2\pi kT}}\, e^{-m_H v^2/2kT} v\, dv\,.$$

Put $u = m_H v^2/2kT$ to obtain

$$\int_0^\infty 2\sqrt{\frac{kT}{2\pi m_H}}\, e^{-u}\, du = 2\sqrt{\frac{kT}{2\pi m_H}}\,.$$

Then $A = 4\pi B^2 n m_H \bar{v}_r/2 = 2B^2 n\sqrt{2\pi kTm_H}$.

C.2 Kinetic energy, escape velocity and redshift

This is the start of the detailed calculations of the energy production.

Throughout the appendix the standard Schwarzschild metric is used

$$(C.3) \qquad\qquad c^2\, ds^2 = -Q c^2\, dt^2 + \frac{1}{Q} dr^2 + r^2\, d\Omega^2,$$

where $Q = 1 - S/r = 1 - 2GM/c^2 r$. Here t is thought of as time, r as radius and $d\Omega^2$, the standard metric on the 2-sphere, is an abbreviation for $d\theta^2 + \sin^2\theta\, d\phi^2$ (or more symmetrically, for $\sum_{j=1}^3 dz_j^2$ restricted to $\sum_{j=1}^3 z_j^2 = 1$ and $S = 2GM/c^2$ is the Schwarzschild radius. Note that $\sqrt{-ds^2}$ can be regarded as proper time.

It is necessary to discuss KE. As remarked earlier, this is not a relativistic concept. It makes sense in Minkowski space where there is the Einstein formula for the KE of a particle of mass m moving with velocity v

$$(C.4) \qquad\qquad mc^2 \left(\frac{1}{\sqrt{1 - v^2/c^2}} - 1 \right)$$

and therefore it makes sense in an inertial frame of reference.

Consider a particle falling freely and radially into a Schwarzschild black hole (and hence following a geodesic). Use τ for proper time along this geodesic. Let $\dot{r}$ denote $dr/d\tau$. The MacKay–Rourke paper [70] describes two natural flat observer fields, the escape field and the dual capture field. Use the latter. This gives a foliation by geodesics following inward freefall paths with orthogonal flat space slices (ie isometric to Euclidean 3-spaces). Thus there are local coordinates with time being proper time along the geodesics and space defined by flat Euclidean coordinates in the orthogonal space slices. These local coordinates provide convenient inertial frames in which to measure KE.

Now the flat slices are derived by making the distance between spheres of area $4\pi r_1^2$ and $4\pi r_2^2$ be $|r_2 - r_1|$ and hence r is a Euclidean coordinate and it follows that $\dot{r}$ is the correct definition of radial velocity for calculating KE. For tangential velocity, θ, ϕ provide standard spherical coordinates in this inertial frame and the usual Euclidean formula for velocity in (r, θ, ϕ) (again measured wrt τ) provides the correct velocity v to measure KE in Equation (C.4).

A formula for escape velocity is also needed. MacKay and Rourke provide this in [70, Equation (10)] namely $\dot{r} = c\sqrt{1 - Q} = \sqrt{2GM/r}$. [MacKay and Rourke use natural units with $G = c = 1$, a factor c has been added to convert to MKS units.]

In the next section these formulae are derived by a simple direct analysis but first here is the promised formula from which the Eddington radius R can be read.

Recall from Section 6.3 that the *Eddington sphere* of radius R is defined by equating outward radiation pressure on the protons in the medium with inward gravitational attraction from the black hole. Also recall the standard equation for the luminosity at the Eddington limit, [73, page 5]

(C.5) $$L_E = \frac{4\pi}{\kappa} GMc$$

where κ is the radiative opacity for electron scattering which is usually taken to be $0.4\text{cm}^2/\text{g}$ or 4×10^{-2} in MKS units [73, page 5]. The Eddington sphere is defined by the same considerations and hence this gives the radiation emitted from this sphere. Note that this formula does not depend on the radius of the radiating sphere. Since it corresponds to local balance of forces, it is true in a relativistic setting provided it is stated exactly where it is applied. It is applied near the Eddington sphere.

Now assume that the luminosity is, within a factor X, the same as the KE of accreted matter falling onto the Eddington sphere. The intuitive description that given in Section 6.3 of the nature of the Eddington sphere suggests that about $1/2$ of the KE released on "impact" should be radiated outwards and about $1/2$ absorbed into the medium below so that X is roughly $1/2$. But, as will be seen later, there is also energy arriving upwards from inside the sphere, and this suggests a larger figure for X. This estimate will be revisited later, but for now keep X as a parameter to be determined.

Equating X times the KE released on impact with the Eddington luminosity gives

(C.6) $$X A c^2 \left(\frac{1}{\sqrt{1 - v^2/c^2}} - 1 \right) = \frac{4\pi}{\kappa} GMc$$

where $v = \sqrt{2GM/R}$ is the escape velocity at R, the velocity of freely infalling matter. Matter does not in fact arrive radially because of tangential motion, which is amplified by conservation of angular momentum as described earlier. However the energy of motion available to be absorbed and re-radiated is unaffected by the transfer of energy from radial to partially tangential and therefore there is no error in assuming that motion is radial here.

It is worth digressing a little here. A particle in the outer region with significant tangential velocity may not reach the Eddington sphere. This happens if the tangential velocity, amplified by conservation of angular momentum, absorbs all the KE and the radial velocity slows to zero. But, because of the mechanics near the Bondi sphere described earlier, particles cannot escape the outer region in significant numbers. It is implicitly assumed that there is a steady state on timescales short compared with that given by the accretion rate. It follows that excess tangential velocity in the outer region must be transmuted into radial velocity by non-thermal particle interaction as suggested earlier. Thus in this region particle interaction allows the plasma to "settle" inwards towards the Eddington sphere, without significant loss of KE. This settling process will need to be modelled in detail in the next version of this work. At this stage just assume that it takes place. There are some features of the process that can be deduced from observations discusssed in Section C.7.

It is not hard to solve Equation (C.6) to find an explicit formula for the Eddington radius R in terms of the other parameters. For calculation purposes however, it is far more convenient to use redshift which has a simple relationship to R. For a black hole with Schwarzschild radius $S = 2GM/c^2$, redshift $1 + z$ at a radius with escape velocity v is $1/\sqrt{1 - v^2/c^2} = 1/\sqrt{1 - S/R}$, since $v = 2GM/R$, and hence $1 - S/R = (1 + z)^{-2}$ or

(C.7) $$S = R(1 - (1 + z)^{-2}).$$

But in terms of z, Equation (C.6) gives the following simple formula for the observed redshift for a black hole radiating from the Eddington sphere:

(C.8) $$z = \frac{4\pi MG}{A c \kappa X}$$

and then substituting for A and B gives:

$$z = \frac{4\pi MG}{2(\frac{2GMm_H}{3kT})^2 n \sqrt{2\pi kTm_H}\, c\kappa X}$$

and collecting terms:

(C.9) $$z = 2^{-1} 9\sqrt{\pi/2}\, \kappa^{-1} M^{-1} n^{-1} (kT)^{1.5} m_H^{-2.5} G^{-1} c^{-1} X^{-1}$$

C.3 Potential and kinetic energy in Schwarzschild space-time

This section gives the promised direct calculation using Schwarzschild geometry for the formulae used in Section C.2 for KE and escape velocity.

Take the approach that a particle is fundamentally described by its 4-momentum, that is, by $P = mU$, where $m = \sqrt{-\langle P, P \rangle}$ is the rest mass of the particle and $U = (\dot{t}, \dot{r}, \dot{\theta}, \dot{\phi})$ is its 4-velocity and dot represents differentiation with respect to proper time.

Consider a particle falling freely in Schwarzschild spacetime, that is following a geodesic path. There are conserved quantities associated to the symmetries of the Schwarzschild spacetime, for example

$$E_0 = -\langle P, \partial_t \rangle.$$

It is tempting to interpret E_0 as the energy measured by a static observer, however this is misleading since ∂_t does not have unit-length and hence does not correspond to a physical observer. There is one exception though, at infinity ∂_t corresponds to an observer comoving with the gravitational source, so it makes sense to interpret E_0 as the energy of the particle measured at infinity by a static observer.

Correspondingly,

$$E := -\langle P, \frac{1}{\sqrt{Q}} \partial_t \rangle = \frac{E_0}{\sqrt{Q}},$$

is regarded as the energy measured by an *interior* static observer, where $Q = 1 - \frac{2GM}{c^2 r}$. Explicitly, $E = iE_0 \sqrt{Q}$.

As the particle falls inwards it gains potential energy

$$PE := E_0 - E = E_0 \left(1 - \frac{1}{\sqrt{Q}} \right)$$

and the relativistic expression for the Kinetic energy can be written as the difference between the observed energy and the rest energy of the particle,

$$KE := E - mc^2$$

and this gives a conservation law of the form

$$KE + PE = E_0 - mc^2$$

where the RHS can be interpreted as the kinetic energy available at infinity. For example, it vanishes when the particle is falling at escape velocity, cf Equation $(C.12)$.

Now elaborate the formula for KE. The proper time parametrisation condition translates to

$$\langle P, P \rangle = -m^2$$

which, for a particle falling radially, reduces to

(C.10) $$-Qc^2\dot{t}^2 + Q^{-1}\dot{r}^2 = -c^2$$

This in turn can be written as a single ODE for r, using the conservation of "energy",

(C.11) $$\dot{r}^2 = c^2 \left(\frac{E_0^2}{m^2c^4} - Q \right).$$

From this it is possible to deduce the escape velocity as measured by proper time. Note that for the particle to get asymptotically to infinity ($\dot{r} = 0$ at $r = \infty$) it is necessary that $mc^2 = E_0$. Hence the velocity necessary to achieve this is

(C.12) $$\dot{r}_{escape} = \pm c\sqrt{1 - Q} = \pm\sqrt{\frac{2GM}{r}},$$

which recovers the classical value.

Remark These geodesics, namely the ones that follow $(\dot{t}, \dot{r}) = (\frac{1}{Q}, \pm c\sqrt{1 - Q})$, are precisely the natural observer fields found by MacKay and Rourke and they correspond to a stream of test particles falling at precisely escape velocity.

Returning to kinetic energy, note that

$$\begin{aligned}
KE &= mc^2 \left(\frac{E}{mc^2} - 1 \right) \\
&= mc^2 \left(\frac{\dot{t}E_0\sqrt{Q}}{mc^2} - 1 \right).
\end{aligned}$$

Dividing (C.10) by $\dot{t}^2$ gives

$$\dot{t} = \sqrt{\frac{Q}{Q^2 - u^2/c^2}},$$

where $u = \frac{\dot{r}}{\dot{t}}$ is the velocity measured by the static coordinates. However, it will be convenient to use the velocity measured by the MacKay–Rourke natural flat observers, that is

$$v = \frac{dr}{d\tau} = \frac{dr}{dt}\frac{dt}{d\tau} = \frac{u}{Q}$$

Therefore the kinetic energy can be written as:

$$KE = mc^2 \left(\frac{E_0}{mc^2 \sqrt{1 - v^2/c^2}} - 1 \right)$$

Note that for the case of a particle falling at escape velocity this reduces to:

$$KE = mc^2 \left(\frac{1}{\sqrt{1 - v^2/c^2}} - 1 \right)$$

C.4   The critical radius and high redshift black holes

Before inserting numbers to compare with observations, there are a couple more pieces of theory. Consider a particle infalling from outside the black hole and suppose that at radius r it releases all its KE, which radiates outwards. The KE is $KE(r) = mc^2(1/\sqrt{Q} - 1)$ where $Q = 1 - 2GM/rc^2 = 1 - v^2/c^2$ and $v = \sqrt{2GM/r}$ the escape velocity at r. The energy $E(r)$ received outside the black hole is $Q = 1/(1+z)^2$ times this, in other words

(C.13) $$E(r) = mc^2(\sqrt{Q} - Q)$$

which is ≥ 0 and zero when $v = 0$ and when $v = c$. The first is natural and obvious but the second is counterintuitive. KE $\to \infty$ as the particle approaches the speed of light at the Schwarzschild radius and you expect the released energy to $\to \infty$ as well. It doesn't.

This mistake occurs in the literature in several places. See for example the discussion in the introduction to [34]. There is no observational difference between a black hole and a super-dense neutron star whose surface is just a little bit above the event horizon. The error is to ignore the redshift reduction in radiated energy.

$E(r)$ has a simple maximum when $Q = 1/4$ so there is a maximum energy released. This depends *only on m and not on M*. Again highly counterintuitive. What does depend on M is the *critical radius* $r = 4S/3$ at which this maximum is achieved. Here $1 - v^2/c^2 = 1/4$ or $v = c\sqrt{3}/2$ and $E(r) = mc^2/4$.

Inside the critical radius the received energy drops off sharply and this allows us to obtain a bound on the radiated energy for black holes whose Eddington radius is $\leq 4S/3$ or equivalently with redshift (calculated at the Eddington sphere) $1 + z \geq 2$ or $z \geq 1$. Let's call these black holes *high redshift black holes*.

The KE for an infalling particle $P(r) = \text{KE}(r) = mc^2(1/\sqrt{Q} - 1)$ represents the maximum energy available to be converted into radiation at that radius, see Section C.3. This conversion as analogous to friction. The medium inside the Eddington radius is "sticky" and slows the particle down, releasing energy. Now normalise so that all radiated energy is measured as received outside the black hole. To do this multiply by $1/(1 + z)^2 = Q$. Assume that the emissions come from inside the critical radius so that the received energy per unit r-distance is decreasing monotonically. Once a portion of $P(r)$ is converted to radiation, it is not replaced, so for maximum effect it needs to be radiated outwards as soon as possible. In other words the maximum possible radiation outwards is obtained by keeping the inward velocity as low as possible (very small KE). So for a bound assume all the KE available at the Eddington radius is radiated outwards and within the Eddington radius set $\dot{r} = 0$ and this gives an upper bound for the extra energy received outside the black hole from below the Eddington radius R:

$$-mc^2 \int_S^R Q \frac{dQ^{-\frac{1}{2}}}{dr} dr$$

$$= mc^2 \int_S^R Q \frac{Q^{-\frac{3}{2}}}{2} \frac{dQ}{dr} dr$$

$$= mc^2 [\sqrt{Q}] \text{ evaluated at } R$$

Since $Q \leq (1/2)\sqrt{Q}$ in this range, this is within a factor 2 of the KE arriving at the Eddington radius from above, and hence the total possible energy radiated outwards is 3 times this KE. In other words, in terms of the notation of Section C.2, it has been proved that $X \leq 3$. However, the assumption that all this energy radiates outwards is unrealistic and the earlier estimate of $X = 1/2$ is much more reasonable.

Note   The same analysis gives a rough upper bound for black holes with small redshift but the result $\sqrt{Q}$ evaluated at the Eddington radius may be far larger than the Eddington luminosity and not provide a useful upper bound. Indeed as $r \to \infty$ it tends to mc^2.

C.5   Calculations

The model will now be compared numerically with observations. In this section various parameters are calculated and, in the next section, their fit with data is tested. MKS units are used throughout, work to 3 significant figures, and use the following constant values:

$\kappa = 4 \times 10^{-2}$, $k = 1.38 \times 10^{-23}$, $m_H = 1.67 \times 10^{-27}$, $G = 6.67 \times 10^{-11}$, $c = 3 \times 10^8$.

Redshift in terms of medium factor and mass

The key equation is the redshift Equation (C.9):

$$z = 2^{-1} 9\sqrt{\pi/2}\, \kappa^{-1} M^{-1} n^{-1} (kT)^{1.5} m_H^{-2.5} G^{-1} c^{-1} X^{-1}$$

For convenience (and familiarity) express M in solar masses; in other words write $M = \mathcal{M} M_{\mathrm{sun}} = 2 \times 10^{30}\, \mathcal{M}$, where $\mathcal{M}$ is the black hole mass in solar masses. Substituting for κ, k, m_H, G, c gives the numerical version which was previewed as equation (6.2):

(C.14) $z = 1.27 \times 10^7\, \mathcal{M}^{-1} n^{-1} T^{1.5}\, [1/(2X)]$

For simplicity use the default value $(\frac{1}{2})$ for X which is the same as ignoring the expression in square brackets. If further information on X comes to light, it can be reinstated.

The factor $n^{-1} T^{1.5}$ depends only on the ambient medium; and is called the *ambient coefficient*, with the notation Θ. Recall that n is the density in particles (protons) per cubic metre and T is the ambient temperature in degrees Kelvin.

The equation now takes the simple form:

(C.15) $z = 1.27 \times 10^7\, \dfrac{\Theta}{\mathcal{M}}$

To get an idea of the range of possible values for Θ, interstellar density is estimated at between 10^2 and 10^{12} where the thinner regions are associated with higher temperatures, which vary inversely with the density from about 10^5 to 10 [24]. Thus Θ varies from about $10^{5.5}$ at the high end (hot thin plasma) to $10^{-10.5}$ at the low end (cold dense gas). [An aside here: "dense" is a relative term. The density of the atmosphere is 10^{25}, and the interstellar density is always far smaller then a laboratory "high vacuum" of about 10^{16}.]

As you can see immediately, the redshift depends critically on the nature of the ambient medium, which can cause it to vary by 16 orders of magnitude. By contrast, the variation with mass, which might be in the range 10^4 to 10^8 solar masses, is far smaller, a further 4 orders of magnitude. For example, given a black hole of mass $10^7 \, M_{sun}$ (a little bigger than SgrA*), so that $10^7 \, \mathcal{M}^{-1} = 1$, then avoiding the extremes for the ambient coefficient, the redshift might vary from 10^{-7}, in other words so small that there is no measurable redshift, up to 10^3 which is so big that the redshift reduction factor in received luminosity, $(1 + z)^{-2}$ or about 10^{-6}, makes it extremely unlikely that the quasar could be detected, unless, like SgrA*, it is very close to us.

Two remarks at this point: (1)   In Chapter 6 it was promised to comment on the maximum density that supports the observed forbidden lines. This is estimated by Greenstein and Schmidt to be about 3×10^{10} [47, third paragraph of abstract] which fits nearly all the densities that have been considered, missing just the extreme cold, dense media.

(2)   It is worth looking at the data for SgrA* since it has just been mentioned. This has mass $4.6 \times 10^6 \, M_{sun}$ and according to the model should have redshift varying from about 10^{-10} to 10^6. A redshift of 10^4 would imply that the received luminosity was 10^{-8} of the Eddington limit, which is exactly what is observed [34, page 1357 top right]. Thus the model suggests that the lack of luminosity for SgrA* is due to a rather hot, thin medium near this black hole.

The data for SgrA* will be examined in detail, at the end of Section C.6.

Three types of redshift and the HLSW formula

The redshift $z = z_{grav}$ used by the model (and quantified above) is the *gravitational* aka *intrinsic* redshift. But when you observe a quasar, you see the *observed* redshift z_{obs} which depends on both the gravitational redshift z_{grav} and the cosmological redshift z_{cos} which is a function of distance.

The relationship between the three is

$$1 + z_{obs} = (1 + z_{grav})(1 + z_{cos})$$

which, provided at least one of z_{grav} or z_{cos} is fairly small, can conveniently be approximated as:

$$z_{obs} \approx z_{grav} + z_{cos}$$

From the cosmological redshift you can read the distance d by the HLSW formula[1] $d = cz_{cos}/H$ where H is the HLSW constant $2.2 \times 10^{-18} \text{sec}^{-1}$. Substituting for c gives:

(C.16) $d = 1.35 \times 10^{26} \, z_{cos}$

The other observed datum is magnitude which is discussed below. From the magnitude and the distance you can calculate the mass. But you need the cosmological redshift, which is not observed, to find the distance. Deciding how to split the observed redshift into intrinsic and cosmological is not simple. The best that can be done is to try various splits and see how they fit. There are however examples (which are referred to as *Arp* quasars) where the observations suggest a galaxy at the same distance as the quasar so that the redshift for this galaxy for z_{cos} can be used.

Specific examples of both these will be looked at in the next section.

Luminosity and magnitude

The main observed data for a quasar are redshift and luminosity, which has a simple relationship to magnitude:

$$L_{obs} = 2.87 \times 10^{-8} \times 10^{-\frac{2}{5}\text{mag}}$$

---

[1]See the convention on page 21.

This is the received luminosity in W/m^2 and the calculation is based on comparison with the solar luminosity ($1.3kw/m^2$) and magnitude (-26.7). In the model, the emitted luminosity is always the Eddington luminosity which depends purely on the black hole mass:

(C.17) $$L_E = \frac{4\pi}{\kappa}GMc = 1.26 \times 10^{31}\mathcal{M}$$

From this you can calculate the received luminosity by applying three correction factors. The first two are straightforward. Use the inverse square law and divide by $1/4\pi$ to convert from total emitted luminosity to received luminosity per unit area and secondly apply redshift correction $(1 + z_{obs})^{-2}$. (If redshifts are small, this second factor can be ignored.)

The third factor is more problematic. Magnitude is usually measured using visible wavelengths, but black hole radiation covers a far wider spectrum. This implies that the observed magnitude underestimates the luminosity by a factor of perhaps 10 or larger. Further the radiation from the black hole is attenuated by intervening clouds for which there is strong evidence (see the discussion in Section C.7) and this gives a further underestimate, which is again difficult to quantify but which might also be up to a factor of 10. Let's call the result of these two the *magnitude correction factor*, denoted Φ, and note that it might vary between 1 and 100 or more.

Thus

$$L_{obs} = \frac{L_E}{4\pi\,\Phi\,d^2(1 + z_{obs})^2}$$

and substituting for the luminosities and distance (using Equation (C.16)), gives the following formula for mass in terms of magnitude and redshifts:

$$\mathcal{M} = \frac{2.87}{1.26}10^{-31} \times 4\pi\,\Phi \times (1.35)^2 \times 10^{52} \times z_{cos}^2(1+z_{obs})^2 \times 10^{-8} \times 10^{-\frac{2}{5}mag}$$

which simplifies to:

$$\mathcal{M} = \Phi \times 5.22 \times 10^{(14-\frac{2}{5}mag)} \times z_{cos}^2(1 + z_{obs})^2$$

To get a feeling for this formula, anticipate the first example in the next section where the data are treated more accurately. Objects 2 and 3 in NGC7603 (see Figure 6.1) both have mag ≈ 20 and $z_{cos} \approx .03$ (taken from the main galaxy) so the formula gives approximately:

$$\mathcal{M} = \Phi \times 5 \times 10^3$$

The gravitational redshift is approx 0.3 and substituting for $\mathcal{M}$ in the redshift formula (C.15) gives:

$$\Phi \approx 10^4 \Theta$$

Thus $\Phi = 1$ (no magnitude correction) corresponds to a black hole of mass about 5×10^3 solar masses floating in a medium of ambient coefficient 10^{-4} which is pretty cold and dense medium. Perhaps the visible filament in which these objects appear to be immersed is a cold dense cloud. Or perhaps, the magnitude correction should be about 100 and the mass 5×10^5, which seems a more likely mass for a quasar, with the medium having a less extreme ambient coefficient of about 10^{-2}.

This section finishes with formulae for the Eddington radius and the temperature of the Eddington sphere (assuming the radiation is black body).

Eddington radius

Recall $1 - S/R = (1 + z)^{-2}$ where S is Schwarzschild radius and R is Eddington radius. Write $\zeta = S/R = 1 - (1 + z)^{-2}$ and notice that for small z, $\zeta = 2z + O(z^2)$. Since the Schwarzschild radius of the sun is 3×10^3 m this gives:

(C.18) $R = 3 \times 10^3 \, \mathcal{M}/\zeta$

Radiant temperature

Suppose the radiation is effectively black body with temperature T_B (notation intended to keep distinct from T which is ambient temperature used earlier). Stefan-Boltzmann gives total luminosity $4\pi R^2 \sigma T_B^4$, where $\sigma = 5.67 \times 10^{-8}$ and equating this with Eddington luminosity gives:

$$4\pi \times 9 \times 10^6 \, \mathcal{M}^2 \times 5.67 \times 10^{-8} \, T_B^4/\zeta^2 = 1.26 \times 10^{31} \, \mathcal{M}$$

which gives:

(C.19) $T_B^4 = 1.96 \times 10^{30} \, \mathcal{M}^{-1} \zeta^2$

Example $\mathcal{M} = 10^6$, $z = 0.1$ so that $\zeta^2 \approx 0.04$ then $T_B \approx 1.67 \times 10^5$.

C.6 Data

Now proceed to examples, that is, given the data z_{cos}, z_{grav} and magnitude use the model to deduce luminosity, mass, ratio R/S, distance to Earth and temperature of the source as if it were a black body.

Continue to use the default value $\frac{1}{2}$ for X and ignore the correction factor Φ (ie assume that it is 1). To take these into account, use the following rules. Multiply Θ by $X/2$ and further multiply both $\mathcal{M}$ and Θ by Φ.

First consider the system around NGC 7603, previewed in the last section, which appears to contain two Arp quasars (objects 2 and 3 in Figure 6.1). Lopez Corredoira and Gutierrez [68] report $z = 0.0295$ and $B = 14.04$ mag for the main galaxy, NGC 7603. A fact that attracted attention is its proximity to NGC 7603B (Object 1 hereafter), a spiral galaxy with higher redshift $z = 0.0569$, moreover a filament can be observed connecting both galaxies. They also found two objects superimposed on the filament with redshifts 0.394 ± 0.002 and 0.245 ± 0.002 for the objects closest to and farthest from NGC 7603, Objects 3 and 2, respectively. B-magnitudes corrected for extinction (due to the filament) are respectively 21.1 ± 1.1 and 22.1 ± 1.1 [68].

They go on to say "If we consider the redshifts as indicators of distance, the respective absolute magnitudes would be : $M_V = -21.5 \pm 0.8$ and -18.9 ± 0.8. However, if we consider an anomalous intrinsic redshift case (in such a case, in order to derive the distance, we set $z = 0.03$), the results are: $M_V = -15.2 \pm 0.8$ and -13.9 ± 0.8 resp. In this second case, they would be on the faint tail of the HII-galaxies, type II; they would be dwarf galaxies, 'tidal dwarfs', and this would explain the observed strong star formation ratio: objects with low luminosity have higher EW(H$_\alpha$). Of course, this would imply that we have non-cosmological redshifts. . . . From several absorption lines we estimated the redshift of the filament apparently connecting NGC 7603 and NGC 7603B as $z = 0.030$, very similar to the redshift of NGC 7603 and probably associated with this galaxy."

This analysis suggests setting $z_{cos} = 0.03$ for the group and $z_{grav} = z - z_{cos}$. Hence the HLSW distance, $d = c \times z_{cos}/H = 13.5 \times 10^{25} \times z_{cos}$, is 4.05×10^{24} metres in this case.

Next, the ratio between the Eddington radius and the Schwarzschild radius is $R/S = 1/1 - (1 + z_{grav})^{-2}$, this gives 18.6, 3.12 and 2.17 for Objects 1, 2 and 3, where z_{grav} has been taken to be equal to 0.028, 0.213 and 0.361, respectively.

The luminosity (in W/m^2 received at Earth) is given in terms of the magnitude by $L_{mag} = 2.87 \times 10^{-8} \times 10^{-\frac{2}{5}mag}$. This gives 5.468×10^{-15}, 5.468×10^{-17} and 7.904×10^{-17} for Objects 1, 2 and 3, respectively.

Obtain the mass by comparing the formulae for the Eddington luminosity and the magnitude luminosity, $\mathcal{M} = M/M_{sun} = 4\pi d^2 L_{mag} \times (1 + z)^2 \times 1.26^{-1} \times 10^{-31}$. Thus $\mathcal{M} = 9.45 \times 10^4$, 1.32×10^3 and 2.39×10^3 for objects 1, 2 and 3, respectively.

The temperature of the quasar as if it were a black body is given by Stefan's law $T_B = \left(L(1 + z)^2/\sigma 4\pi R^2\right)^{\frac{1}{4}}$ and in terms of previous data it is

$$T_B = \left(L_{mag}(1 + z)^2 \times 1/\sigma \times d^2 \times (S/R)^2 \times (1/\mathcal{M})^2 \times (1/S_{sun})^2\right)^{\frac{1}{4}}.$$

For Objects 1, 2 and 3 this gives 4.95×10^5, 3.52×10^6 and 3.63×10^6, respectively.

Finally, the ambient coefficient is defined by $\Theta = 10^{-7} z \mathcal{M}$, which helps to constrain the possible values of the ambient density and temperature. For the case at hand this gives 2.07×10^{-4}, 2.19×10^{-5} and 6.75×10^{-5} for objects 1, 2 and 3, resp.

A spreadsheet has been used for these calculations, and the results for these and several more examples, are in the tables which follow. Included are two quasars (3C273 and 3C48) for which the redshift split is unknown and for which various splits have been tried. The examples come from Galianni, Arp, Burbidge, et al [45], Lopez Corredoira and Gutierrez [67, 68], Greenstein and Schmidt [47], and Hoyle and Burbidge [53].

Lopez Corredoira-Gutierrez

	INPUTS						OUTPUTS					
	z			Magnitude	R/S	L_{mag}	Solar masses	Distance	T_B	$T_B * 1/1+z$		Ambient coefficient
	Obs	Cos	Grav			W/m^2					X	$n^{-1} * T^{1.5}$
NGC 7603	0.029	0.03	0	14.04	-	6.948E-14	1.136E6	4.050E24	-	-	-	-
Object 1	0.058	0.03	0.028	16.8	1.861E1	5.469E-15	9.449E4	4.050E24	4.951E5	4.816E5	0.5	2.067E-4
Object 2	0.243	0.03	0.213	21.8	3.121E0	5.469E-17	1.316E3	4.050E24	3.519E6	2.901E6	0.5	2.189E-5
Object 3	0.391	0.03	0.361	21.4	2.173E0	7.905E-17	2.394E3	4.050E24	3.631E6	2.668E6	0.5	6.752E-5
NEQ 3												
Object 1	0.1935	0.12	0.0735	19.8	7.562E0	3.450E-16	1.040E5	1.620E25	7.582E5	7.063E5	0.5	5.973E-4
Object 2	0.1939	0.12	0.0739	19.6	7.525E0	4.148E-16	1.252E5	1.620E25	7.257E5	6.758E5	0.5	7.226E-4
Object 3	0.2229	0.12	0.1029	20.2	5.621E0	2.387E-16	7.596E4	1.620E25	9.513E5	8.625E5	0.5	6.107E-4
Object 4	0.1239	0.12	0.0039	17.3	1.290E2	3.450E-15	9.097E5	1.620E25	1.068E5	1.064E5	0.5	2.772E-4
GC 0248+430	0.051	-	-	-								
QSO 1	1.311	0.051	1.26	17.45	1.243E0	3.005E-15	7.253E5	6.885E24	1.151E6	5.091E5	0.5	7.140E-2
QSO 2	1.531	0.051	1.48	21.55	1.194E0	6.885E-17	2.001E4	6.885E24	2.881E6	1.162E6	1.5	6.940E-3
B2 1637+29	0.086	-	-	-								
Partner	0.104	0.086	0.018	-								
Aligned QSO	0.568	0.086	0.482	20	1.836E0	2.870E-16	8.470E4	1.161E25	1.620E6	1.093E6	1.5	9.568E-3

Hoyle–Burbidge, Arp–Burbidge et al

	INPUTS						OUTPUTS					
	z			Magnitude	R/S	L_{mag}	Solar masses	Distance	T_B	$T_B * 1/1+z$		Ambient coefficient
	Obs	Cos	Grav			W/m^2					X	$n^{-1} * T^{1.5}$
NGC 4319	0.0057	0.0057	0	-	-	-	-	7.695E23	-	-	-	-
MK 205	0.07	0.0057	0.0643	14.5	8.534E0	4.549E-14	3.041E4	7.695E23	9.706E5	9.120E5	0.5	1.528E-4
NGC 3067	0.0047	0.0047	0	-	-	-	-	6.345E23	-	-	-	-
3C 232	0.533	0.0047	0.5283	15.8	1.749E0	1.374E-14	1.288E4	6.345E23	2.658E6	1.739E6	0.5	5.314E-4
ESO 1327-2041	0.018	0.018	0	-	-	-	-	2.430E24	-	-	-	-
QSO 1327-206	1.17	0.018	1.152	16.5	1.275E0	7.209E-15	1.965E5	2.430E24	1.575E6	7.318E5	0.5	1.769E-2
Gal 0248+430	0.051	0.051	0	-	-	-	-	6.885E24	-	-	-	-
Q 0248 +430	1.1311	0.051	1.0801	17.45	1.301E0	3.005E-15	6.144E5	6.885E24	1.173E6	5.638E5	0.5	5.185E-2
Gal Abell 2854	0.12	0.12	0	-	-	-	-	1.620E25	-	-	-	-
2319+272 (4C 27.50)	1.253	0.12	1.133	18.6	1.282E0	1.042E-15	1.240E6	1.620E25	9.911E5	4.646E5	0.5	1.098E-1
NGC 3079	0.00375	0.00375	0	-	-	-	-	5.063E23	-	-	-	-
0958+559	1.17	0.00375	1.16625	18.4	1.271E0	1.253E-15	1.502E3	5.063E23	5.336E6	2.463E6	0.5	1.368E-4
Arp, Burbidge, et al.												
NGC 7319	0.022	0.022	0	-	-	-	-	2.970E24	-	-	-	-
QSO	2.114	0.022	2.092	21.79	1.117E0	5.519E-17	4.640E3	2.970E24	4.293E6	1.388E6	0.5	7.583E-4

Greenstein–Schmidt

	INPUTS						OUTPUTS						
	z			Magnitude	R/S	L_{mag}	Solar masses	Distance	T_B	$T_B * 1/1+z$		Ambient coefficient	Spectral index
	Obs	Cos	Grav			W/m^2					X	$n^{-1} * T^{1.5}$	
3C 273	0.1581	0.001	0.1571	12.6	3.951E0	2.617E-13	4.768E3	1.350E23	2.437E6	2.106E6	0.5	5.852E-5	0.9
	0.1581	0.01	0.1481	12.6	4.143E0	2.617E-13	4.768E4	1.350E24	7.496E5	6.529E5	0.5	5.517E-3	0.9
	0.1581	0.05	0.1081	12.6	5.388E0	2.617E-13	1.192E7	6.750E24	2.888E5	2.606E5	0.5	1.007E-1	0.9
	0.1581	0.1	0.0581	12.6	9.363E0	2.617E-13	4.768E7	1.350E25	1.514E5	1.431E5	0.5	2.164E-1	0.9
	0.1581	0.158	1E-04	12.6	5.001E3	2.617E-13	1.190E8	2.133E25	5.066E3	5.066E3	0.5	9.299E-4	0.9
3C 48	0.3675	0.001	0.3665	16.2	2.153E0	9.503E-15	3.224E2	1.350E23	6.022E6	4.407E6	0.5	9.231E-6	1.25
	0.3675	0.01	0.3575	16.2	2.187E0	9.503E-15	3.182E4	1.350E24	1.896E6	1.397E6	0.5	8.886E-4	0.95
	0.3675	0.05	0.3175	16.2	2.359E0	9.503E-15	7.492E5	6.750E24	8.286E5	6.290E5	0.5	1.858E-2	0.95
	0.3675	0.1	0.2675	16.2	2.649E0	9.503E-15	2.774E6	1.350E25	5.638E5	4.448E5	0.5	5.797E-2	0.95
	0.3675	0.2	0.1675	16.2	3.754E0	9.503E-15	9.413E6	2.700E25	3.489E5	2.988E5	0.5	1.232E-1	0.95
	0.3675	0.367	0.0005	16.2	1.001E3	9.503E-15	2.328E7	4.955E25	1.704E4	1.703E4	0.5	9.093E-4	0.95

Finally consider data for SgrA*. According to [34], the received luminosity is 1.85×10^{-13} W/m^2 which is approximately 10^{-8} of the Eddington limit.

Accordingly set $z_{\text{grav}} = 10^{-4}$, which gives the following data in the same format as above.

Sgr A* data

	z		R/S	L_{mag}	Solar masses	Distance	T_B	$T_B * 1/1+z$		Ambient coefficient
Obs	Cos	Grav		W/m^2	$\mathcal{M}$				X	$n^{-1} * T^{1.5}$
10^4	0	10^4	100	1.185E-13	4.300E6	2.592E20	5.269E5	5.268E1	0.5	3.516E3

This table predicts the observed temperature for Sgr A* of about 50 K, which fits well with observations in the radio frequency range. The spectrum of Sgr A* from Narayan–McClintock [78, page 6] is reproduced in Figure C.1.

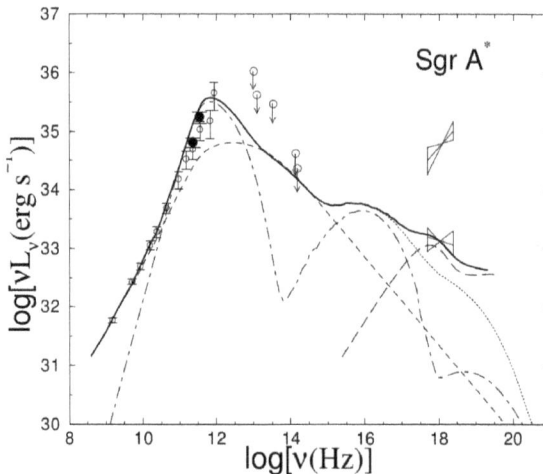

Figure C.1: Figure 3 from [78] where the following references can be found. The radio data are from Falcke et al (1998; open circles) and Zhao et al (2003; filled circles), the IR data are from Serabyn et al (1997) and Hornstein et al. (2002), and the two "bow-ties" in the X-ray band correspond to the quiescent (lower) and flaring (higher) data from Baganoff et al (2001, 2003).

Ignoring the solid and dotted lines (which are attempts to fit the data with current models), the radio frequency observations and infra-red observations (up to about 10^{14} Hz) are a pretty good fit for a black body radiator with peak output at about 5×10^{12} Hz which corresponds to a temperature of about 50 K (see the frequency-dependent formulation of Wien's law in [30]) and fits the data well. Note that the actual temperature of the Eddington sphere is 5×10^5 K; it is the apparent temperature, after redshift adjustment, which is 50 K. The extreme redshift of $\mathrm{Sgr\,A^*}$ explains why the principal radiation falls in the radio frequency range. The two "bow-ties" are probably due to activity remote from the actual black hole, perhaps associated with orbiting clouds in the outer region. This illustrates clearly that the model is merely a first approximation to reality, applying only to the main black hole radiator, and omits other important features.

C.7 Conclusions

Chapter 6 and this appendix has investigated a very simple model for black hole radiation which appears to explain the observations of Arp and the paper of Hawkins [52], both of which suggest that quasars typically exhibit redshift that is not cosmological.

It is not suggested that the model is a perfect fit for all the facts. One obvious set of data that need a more complicated model are the Spectral Energy Distributions (SEDs) for quasars which are typically quite complicated and far from simple black body graphs; for a fairly simple example see Figure C.2 right. By contrast, the composite spectrum on the left does have the rough outline of a black body, suggesting that the basic mechanism for

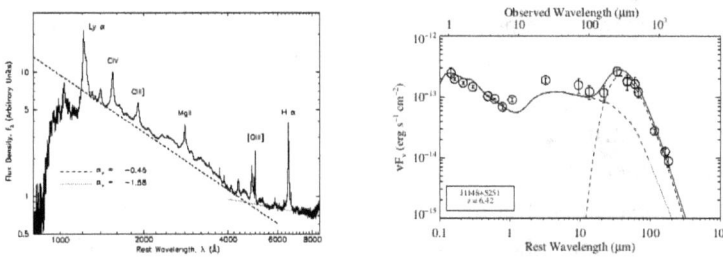

Figure C.2: Left: composite spectrum (figure 3 from [100])   Right: spectrum of the $z = 6.42$ quasar SDSS J1148+5251 (figure 1 from [63])

radiation is by thermal excitation, as in the model. One obvious suggestion for correcting SEDs is to take into account the orbiting clouds, responsible for the observed variation in radiation and which absorb radiation. The spectrum on the right could plausibly result from a black body spectrum which is partially obscured causing the two dips at the top. Or perhaps, like Sgr A* there is a black body radiator in the longer wavelengths with some short wavelength activity from the outer region superimposed.

Another strong piece of evidence (apart from variability) for the existence of orbiting clouds is the so-called "Lyman-alpha-forest". The clouds on the path to us cause absorption lines and the principal line is the L_α-line. The clouds are all at different redshifts and these lines form a forest, see Figure C.3. The existence of the L_α-forest is used by Wright [105] to prove

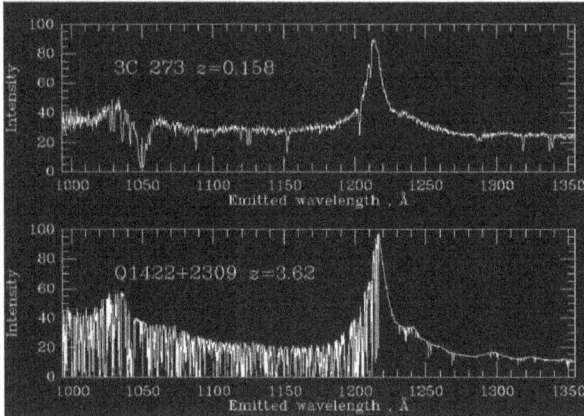

Figure C.3: The Lyman Alpha Forest at low and high redshift, taken from [105]

(fallaciously) that Arp is wrong about intrinsic redshift. He assumes that if the redshift is intrinsic then it jumps down suddenly away from the quasar and therefore there should be a gap to the left of the main L_α emission line before the forest starts. But the absorption clouds can orbit as close as they like to the Eddington sphere, and there is no reason for there to be a gap.

The L_α-forest suggests strongly that the settling process, that is hypothesised to take place in the outer region, tends to form strata. This is plausible because once a stratum of greater density starts to build up, then interaction with other particles becomes more likely, and this will often result in material added to the stratum. This is analogous to the instability observed in many queuing or draining situations (for example traffic congestion with most of the traffic locked up in stationary bands at any one time). These strata are responsible both for the observed L_α-forest and the quasar variability. As remarked earlier, the outer region needs proper modelling, and the authors intend to return to this in a later paper.

However, there are complicated features for many quasars which are not adequately explained by the simple model exposited in this appendix, even with added absorption clouds and strata. For heavier quasars, whose redshift is largely cosmological, the current theory is probably much more appropriate, especially when there are features such as jets which can be observed. It is only suggested that the theory fits smaller black holes with high intrinsic redshifts, which are probably much smaller and closer than current theory suggests. Note that very high redshift examples are very dim because of the redshift reduction in energy received and therefore unlikely to be observed.

Appendix D

Local stellar velocities

As remarked in Chapter 8, there has been a huge effort expended mapping the velocities of stars in our neighbourhood. There are some (apparently) paradoxical properties of these excellent observations which all have natural explanations in the new model exposited in this book. These (fairly technical) explanations are given in this appendix.

D.1   The observations

The discusson is based on the excellent treatment in Binney and Merrifield [35, Section 10.3]. The first and most important point that must be understood is that the observations are all *relative to the Sun*. There is no way of determining absolute motion (eg with respect to the centre of the galaxy) from these observations. If a model for galactic motion is chosen (eg the current conventional model of roughly circular motion in the plane of the galaxy) then absolute motion can be deduced, but other models give other results.

The coordinate system used to express observations is (x, y, z) where x points from the Sun to the centre of the galaxy, y is perpendicular to x in the plane of the galaxy and points roughly in the direction that the Sun is moving and z is perpendicular to both and points to the galactic north pole.

By convention, velocities in these three directions are denoted U, V, W respectively.

The salient features of the observations are:

(1) The Sun is moving with velocity $(U, V, W) \approx (10, 5, 7)$km/sec with respect to the average velocity of nearby stars.

This velocity is well within the observed variations for stellar velocities for all types of stars in our neighbourhood and therefore this observation is completely unremarkable, unlike the remaining ones. Note that this does *not* imply that the Sun is moving towards the centre ($U > 0$) but merely that its velocity measured with respect to the average velocity for nearby stars has a component towards the centre. In the model presented in this book, stars are moving around the galaxy at the usual tangential velocity of about 200km/sec and also outwards at perhaps 20km/sec, so the Sun is also moving outwards at perhaps 10km/sec.

The remaining observations concern the statistics of the observed velocities for subsets of stars of a given stellar type. The main variable considered is colour "B–V" which for Main Sequence stars is largely determined by age (or rather by metalicity, which for stars in our neighbourhood is inversely correlated with age, see Section 8.2). The reddest observations are ignored to improve the correlation with age, see the comments at the top of page 630 of [35].

(2) The average velocity of Main Sequence stars in our neighbourhood *decreases monotonically* with respect to age.

(3) The variation in velocities (measured for example as the square of the standard deviation of the velocities from the mean velocity) *increases monotonically* with respect to age.

For details here see [35, Figures 10.10, 10.12].

These observations are very remarkable. At first sight there is no reason at all to expect any dynamic properties of stars in the galaxy to depend systematically on age. The two observations can be combined to give a linear relation between velocity and variation, which is called *asymmetric drift*: for all types of stars, velocity decreases linearly with respect to squared variation in velocity [35, Figure 10.11, page 628].

Now consider the variation in velocity as a function of direction. To first approximation, squared standard deviation can be modelled as a quadratic form. This is the so-called *velocity ellipsoid* [35, Box 10.2]. The distance of the ellipsoid surface from the origin in a given direction gives the standard deviation for velocities in that direction. The principal axes of this ellipsoid give intrinsic directions related to the velocity variation. As expected from symmetry considerations, for all types of stars one of the principal axes is parallel to the z-axis (ie towards the galactic north pole) and this is the shortest principal axis. The other two lie in the galactic plane. For the current model of galactic motion in which stars are supposed to move in roughly circular orbits, the x-axis should be a line of symmetry. The final, and most remarkable of these observations is that this is not the case. The major axis of the velocity ellipsoid lies in the galactic plane and points, not towards the galactic centre, but makes an non-zero angle with the x-axis on the side of the positive y-axis of between 10 and 30 degrees approximately. This non zero angle is called *vertex deviation*. The final and most remarkable observation is the following.

(4)   Vertex deviation *decreases* with stellar age.

D.2   The explanations: Velocity variation increases with age

Recall that stellar systems form in the spiral arms by condensation of the background gas stream, together with dust and contaminants from supernova explosions etc, see Section 8.2.

At birth, a star's velocity will be much the same as the average velocity of the gas stream, but once born it is subject to various gravitational forces of a random character from nearby stars and groups of stars and its velocity tends to vary from average in a statistical sense. Thus the older a star is, the longer time it has to acquire random variations and the more variation you would expect. This is observation (3).

D.3   Asymmetric drift

Variations in velocity are mostly due to interactions between nearby stars and groups of stars. Therefore they conserve kinetic energy. When uniform velocities vary randomly from a common average preserving kinetic energy then average velocity *decreases* with the average decrease proportional to the average squared deviation. This explains asymmetric drift and observation (2) follows from observation (3). This is a well-known phenomenon and proved in for example [36, Section 4.2.1]. Here is an elementary proof which gives the dependence on average velocity explicitly.

Assume for simplicity that there is a group of N stars of equal mass all travelling with the same velocity vector $\mathbf{v}$ subject to small random changes preserving kinetic energy. Let the new velocity of the ith star be $\mathbf{v} + \mathbf{e_i}$ then conservation of kinetic energy gives:

$$\sum_i ||\mathbf{v} + \mathbf{e_i}||^2 = \sum_i ||\mathbf{v}||^2$$

which implies

$$\sum_i 2\mathbf{v}.\mathbf{e_i} = -\sum_i ||\mathbf{e_i}||^2$$

divide both sides by $2Nv$ where $v = ||\mathbf{v}||$ and the left hand side becomes the average increase in velocity in the $\mathbf{v}$ direction (negative and therefore a decrease) and the right hand side is $-1/(2v)$ times the average squared variation. It will be seen shortly that, after correcting for the effect of inertial drag (replacing $\mathbf{v}$ by $\mathbf{v}_{inert}$ see Figure D.1), the principal source of velocity variation is roughly in the $\mathbf{v}$ direction and therefore the major change of velocity is roughly parallel to $\mathbf{v}$ and hence the average velocity decrease is proportional to average squared variation with the constant of proportionality being $1/(2v)$.

Comparing this with [35, Equation 10.12] gives $2v = 80\,\mathrm{km/sec}$ and hence $v = 40\,\mathrm{km/sec}$ approximately. This is not the observed velocity for the Sun against distant objects which, in common with all observed rotation curves, is approximately 200km/sec. To explain this discrepancy, which is caused by inertial drag, it is necessary to recall the analysis of Chapter 5. Write the

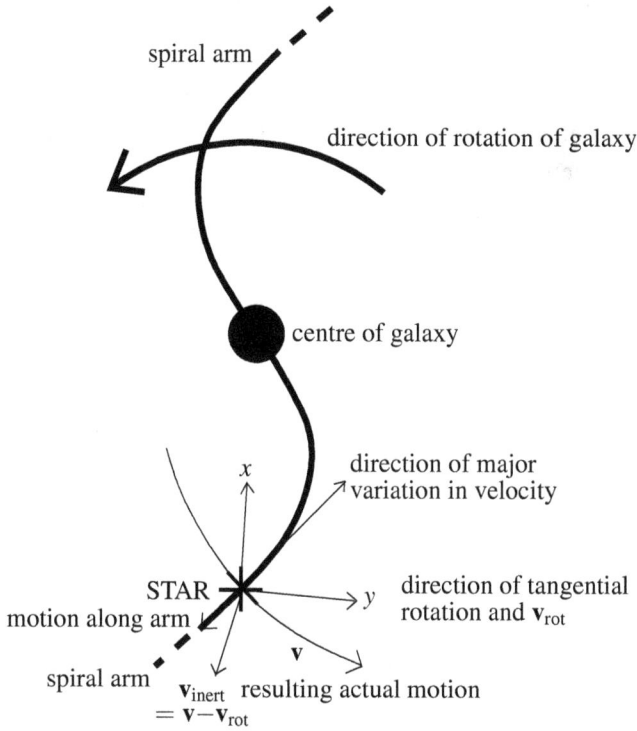

Figure D.1: The velocities near a star

velocity vector of a particle moving in the plane of the galaxy as the sum:

$$\mathbf{v} = \mathbf{v}_{rot} + \mathbf{v}_{inert}$$

where $\mathbf{v}_{rot}$ is the velocity due to rotation of the local inertial frame and $\mathbf{v}_{inert}$ is the velocity measured in the local inertial frame. Note that the notation here is not the same as used in Chapter 5, where v was tangential velocity and not total velocity. The use of bold face is intended to make this distinction clear. Now the conservation of energy applies only to $\mathbf{v}_{inert}$ and it is twice the size of this velocity which is the inverse of the constant of proportionality. Tangential velocity is mostly rotational near the centre and moving outwards, the inertial part of tangential velocity grows asymptotically to a maximum of half the asymptotic limit of 200km/sec. Radial velocity is all inertial but decreases outwards as you would expect. The nett effect is that an inertial volocity on average of size roughly 40km/sec is consistent both with the model and with observations.

D.4   Vertex deviation

To understand vertex deviation it is necessary need to think carefully about
the geometry that was analysed in Chapter 7, which produces the classic
spiral structure. Consider Figure D.1; a star moves tangentially with the
rotating galaxy and also outwards along the arm in which it lies. The nett
effect is a spiral in the *opposite* direction as illustrated.

Some preliminary remarks are needed. The variation in velocity is a relative
effect and depends only on the interaction of stars in the frame moving with
a star. Therefore it is the apparent arrangement of stars which is important,
in other words the visible spiral structure, called the *spiral frame*, in which
stars move along the arms. Further the direction of variations is preserved
as the stars move in the spiral frame.

Now the main source of random variations in stellar velocities is from the
fact that the arms, created as they are by a series of explosions in the belt,
are not uniform. Hence the component of velocity *along* the arm is subject
to the major variation. But this is in a direction towards the centre near the
root of the arm and then turns away in a direction towards the direction of
rotation as illustrated in Figure D.1. Thus the major variation is not towards
the centre (along the x-axis) but has a component along the y-axis which
is precisely the observed vertex deviation. But by inspecting the shape of
the arm, it can be seen that the younger a star is, the greater will be the
proportion of its life spent in the outer region of the arm, where the direction
of variation is further from the centre and hence the greater will be the vertex
deviation. This is observation (4).

Appendix E

Optical distortion in the Hubble Ultra-Deep Field

E.1   Introduction

The Hubble Ultra-Deep Field (HUDF) [8] provides a unique snapshot of the universe at a great distance (and hence time) removed from our immediate neighbourhood. There are many strange looking galaxies in the field and the purpose of this appendix is to examine a selection of these galaxies and to suggest that their strange appearance is not instrinsic but rather due to optical distortion caused by non-uniformity in the intervening space-time, and that the galaxies being viewed are in fact similar to a field of comparable size in a closer neighbourhood.

Patterns of non-uniformity in space-time are usually called "gravitational waves", which expresses graphically the way that they propagate with respect to a particular time parameter and this terminology will be used frequently. Now one of the main hypotheses of this book is that big spiral galaxies are rotating and in so doing they create inertial drag fields which propagate at the speed of light. This implies that the universe is filled with low level gravitational disturbance, and therefore the effects of this are expected to be seen. There are also gravitational disturbances coming from movements of heavy objects other than rotation and indeed natural observer fields which are also associated with heavy objects also have a distorting effect on space-time. (This is used in the explanation for redshift in Section 9.4.)

E.2   The face galaxy

The discussion starts by examining the clearest example and one where it is possible to describe a simple gravitational field which produces the visible distortion. This is the "face galaxy" copied in Figure E.1.

Figure E.1: The face galaxy

Note   You are recommended to download a copy of the highest resolution jpeg of the HUDF as instructed in the bibliography at [8]. To help you find a particular galaxy or image instrinsic coordinates are given from the bottom left, where the height and width are 1 unit and coordinates are taken mod 1 (so that a negative number is a coordinate from the right or top). The face galaxy is at $(.42, -.09)$.

If the face galaxy is an accurate representation of a real galaxy, then it is one of the weirdest galaxies you can imagine. It has two centres. They must be in the process of merging. A far more chaotic structure would be expected from such a merger and moreover there is no reason at all to expect the colours to match so accurately. Far more plausible is that the two centres are the same and that the appearance is due to some kind of optical reflection process. Looking more closely, there is a rough line of symmetry in the centre (marked with dashes in Figure E.2).

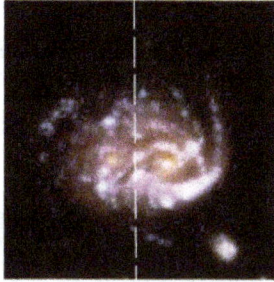

Figure E.2: The rough line of symmetry

The symmetry is near perfect in the top half (near the line of symmetry) and not so accurate in the bottom. So apart from this reflection, there is some other distortion going on. Looking carefully at the line of symmetry, there are some white dots as it crosses some of the denser parts of the galaxy. If the reflection is due to a lensing effect then there will be an element of focussing at the line of reflection and this will produce a bunching of light paths and explain these white dots (more detail on this will be given below). The symmetry breaks down at the outside where there are clear spirals going the same way and not mirror images, but now that it is known how to recognise a mirror line then another slightly slanting to the left (dashed in Figure E.3) can be seen.

Figure E.3: The second mirror line

Finally, fold along these mirror lines and cut out the middle (and the spurious white dots) and paste the outsides together. This has been done on the right in Figure E.4. On the left in Figure E.4 is the original galaxy with the two mirror lines dashed and the two cut lines (which coincide after both reflections) shown solid. The final picture on the right in Figure E.4 is obtained by cutting along the cut lines, discarding the middle and pasting the two outside pieces together. It is close to a standard spiral galaxy (with just a little residual distortion).

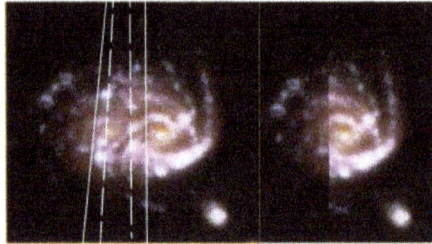

Figure E.4: Cutting and gluing

E.3   Gravitational solitons

Before examining other funny objects in the HUDF it is worth pointing out that there is a simple gravitational field which produces exactly the distortion seen in the face galaxy (reflection in two roughly parallel mirrors) namely a *gravitational soliton*.

The description of this field is in terms of distortion of the metric on space and ignores the accompanying distortion in time. This is justified since the spatial distortion is small and relativistic effects minimal. A proper treatment would treat both space and time.

Suppose given two concentric spheres of fairly large radius with a relatively small gap between them. Suppose that the metric on the gap is altered so that radial distance is changed by a fixed scale factor close to 1, tangential distance being unaltered. If the factor is greater than 1, this is called a *positive* soliton and if the factor is less than 1, a *negative* soliton. It is not hard to describe the geodesics in this metric. Outside the gap, they are of course straight lines. In the gap they are circles. This is easy to see for the positive case where a plane section through the centre is isometric to a portion of a cone, which can be flattened and the geodesics drawn. In this case the circles are concave towards the centre of the spheres. In the flat case the geodesics in the gap are straight lines and, by extrapolation, in the negative case, they are circles concave outwards. When a geodesic crosses one of the spheres it makes an apparent bend, namely the tangent of the angle to the tangent plane is scaled by the same factor as the metric scale. The bend is describes as "apparent" because the geodesic is straight as it crosses the sphere if the local metric scaling is performed.

Now suppose that there is a negative gravitational soliton between us and the face galaxy with a tangent plane passing through our eye and the galaxy. It can be seen that the image of the face galaxy has two roughly parallel mirrors. Look at Figure E.5.

Three typical light paths from our eye to the galaxy have been drawn. Path 1 is straight and panning left, moving in from the right, paths stay straight until they reach tangency to the outer sphere. At this point they start to contain a portion of a circle which is concave to the right and causes the paths to bend to the right as typified by path 2. This bending increases (and the far end of the path pans to right) until tangency to the inner sphere is reached, when the path becomes three straight lines with two smaller circular portions as typified by path 3. The paths now continue to pan to the left. Thus there are two places where movement of the far ends of the paths reverses and this gives the double mirror effect.

Robert MacKay points out that a mirage has a similar mechanism and may be more familiar than a gravitational soliton!

Finally notice that at the points of reflection there will be a focussing effect. The metric described is not C^∞ but merely C^1. If a C^∞ approximation is used then there is a non-zero angle of paths all roughly converging to the same point at the reversal times and this gives rise to the white blobs seen on the miror lines in the face galaxy (assuming that the mechanism at work in the face galaxy is similar to the one described here).

GALAXY

soliton

1

2

3

EYE

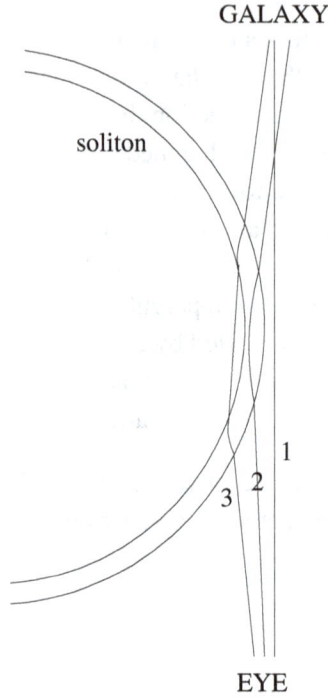

Figure E.5: The soliton in action

E.4   The companion face

At (.40, .50) there is a very similar object, Figure E.6 (left) The similarity is more apparent if it is rotated, Figure E.6 (right).

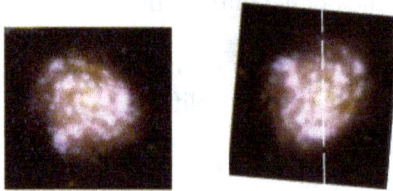

Figure E.6: The companion face (left) and rotated (right)

Now there is a clear (and very rough) vertical line of symmetry (marked dotted) but there the analogy with the face galaxy stops. It is difficult to finish the description of the precise distortion that must have happened to make a standard spiral galaxy look like this. But it is clear that this is again a distorted spiral galaxy.

The colouring is very similar to the original face (Figure E.1) and it is just possible that both these two galaxies are two distorted images of the same galaxy.

E.5   The group of four

At (.39, −.16) is a group of four galaxies: two "white" and two "orange", Figure E.7.

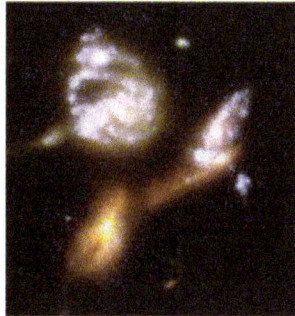

Figure E.7: The group of four

The left-hand white galaxy is clearly an ordinary spiral galaxy showing optical distortion: the centre has been elongated (top-left to bottom-right) and, to the left and top, there is a pair of spiral arm sections which have been dragged out; they look as though they are on a sheet which has been bent up. The other white galaxy is severely distorted with a clear sloping "cut-off" plane to the left. This would be due to a planar gravitational wave front in the intervening space. Moreover these two galaxies have very similar colour and light distribution and most probably they are in fact two images of the same galaxy. The "reflection" plane would be associated with the same wave front that is causing the cut-off in the right-hand image.

The two orange galaxies are both severely distorted and again are quite likely to be different images of the same galaxy.

E.6   Four distorted spirals

In Figure E.8 are four galaxies from different parts of the field. Their coordinates are $(.31, -.16)$, $(.13, -.33)$, $(-.09, -.14)$, $(.12, -.24)$ respectively.

Figure E.8: Four distorted spirals

Each is a spiral galaxy with optical distortion. On the left is a galaxy having a "bad hair day" caused by image distortion on the right-hand side. Middle-left is a spiral galaxy with anomalous straight section in one arm (top left). Although this could plausibly be an undistorted image, it seems more likely, given the distortion that seen elsewhere, that this straight section is caused by focussing at a wave front in the intervening space. Middle-right is a distorted spiral with several different kinds of distortion and to the right is a spiral with quite simple distortion causing a "toothpick" appearance.

E.7   Miscellanea

Finally in Figure E.9 is a collection of miscellaneous objects from various parts of the field. The coordinates are (top row) $(.04, .44)$, $(-.04, -.12)$, $(-.25, .09)$, $(.24, .32)$ and (bottom row) $(.14, .37)$, $(.19, .29)$.

Top row left and centre-left are two sets of possibly repeated images of the same object. Top row centre-right (the blue ring galaxy) is probably a highly distorted image of a regular spiral with the ring being a distorted arm with a similar distortion to the left-hand white galaxy in Figure E.8. This galaxy is probably a long way behind the regular edge-on spiral to the left and not interacting with it. Top-row right is a toothpick galaxy, a more extremely

Figure E.9: Miscellanea

distorted (and distant) version of the right-hand galaxy in Figure E.8. The bottom row shows two collections of distorted fragments, which could both be images of the same galaxy or pair of galaxies.

The images in Figure E.9 are typical of many other images in the field. There is a collection of "tadpole" galaxies from the field on the Hubble site (search tadpole) similar to the toothpick galaxies given above, and there are collections of fragments like bottom images all over the field.

One final remark. Most of the distant objects in the field show repeated white dots similar to those found on the mirror line in Figure E.2. These probably have a similar origin in local focussing effects in the distorting gravitational fields between us and these distant objects. For example the ring in Figure E.9 (top centre-right) is probably the image of a fairly smooth arm of a regular spiral with focussing effects causing the grainy appearance.

E.8 Conclusion

All the strange shapes and unfamiliar objects in the HUDF can be explained as optically distorted images of familiar galaxies. Given the clear evidence of such distortion in the field, there are no grounds for concluding that an undistorted view of the universe in the region covered by the field would be qualitatively different from a more local region.

Appendix F

Gamma Ray Bursts

This appendix reproduces a short version of the joint paper with Robert MacKay [72].

A kinematic explanation for gamma-ray bursts

ROBERT S MACKAY AND COLIN ROURKE

Abstract Gamma-ray bursts are flashes of gamma-rays lasting from milliseconds to a few minutes, which then soften progressively to X-rays and ultimately to radio waves. They are observed from all directions in space, roughly uniformly. They have been attributed to cataclysmic events. We propose, however, that many of them may be optical illusions, simply the result of our entry into the region illuminated by a continuously emitting object. At such an entry, the emitter appears infinitely blue-shifted and infinitely bright. We demonstrate the phenomenon in de Sitter space, where much can be calculated explicitly, and then extend the idea to more general space-times.

Keywords Gamma-ray bursts; kinematic effect; de Sitter space

PACS codes 98.70.Rz: Gamma ray bursts, 98.62.Py: Distances, redshifts, radial velocities, 04.20.Jb: Classical general relativity — exact solutions

F.1   Introduction

Gamma-ray bursts were first observed in 1967 during monitoring of the nuclear test ban treaty, but were subsequently realised to come from outside our solar system, indeed outside our galaxy. Dedicated instruments have now detected and continue to detect many of them. There is a highly developed theory of their origins in various types of cataclysmic event, such as collapse of a high-mass star to a neutron star, or capture of a star by a black hole. For a review, see [74].

We propose, however, that many gamma-ray bursts may be optical illusions. If space-time is geodesically complete but an emitting object does not illuminate the whole of space-time, then on our entry into the illuminated region we see the emitter infinitely blue-shifted and infinitely intense. Both the blue-shift and intensity fall off with receiver time. This produces an effect qualitatively similar to the observations of gamma-ray bursts.

We believe the effect has been ignored so far because of Weyl's postulate [102] and the subsequent standard assumption that all matter moves along the HLSW flow[1] in a big-bang Friedmann universe. It can occur, however, in Friedmann universes if they have infinite past and emitter and receiver are not both on the HLSW flow.

We first demonstrate the phenomenon in de Sitter space, where much can be calculated explicitly. Then we extend the idea to more general space-times. Details are given in [72].

F.2   Geodesics in de Sitter space

De Sitter space $\mathcal{DS}$ is the Lorentzian manifold given by restricting 5-dimensional Minkowski space $\mathcal{M}^5$ with metric

$$ds^2 = -dx_0^2 + \sum_{i=1}^{4} dx_i^2$$

---

[1]See the convention on page 21.

to the hyperboloid

$$-x_0^2 + \sum_{i=1}^{4} x_i^2 = R_{DS}^2.$$

The constant R_{DS} is called the de Sitter radius. The metric g on $\mathcal{DS}$ satisfies Einstein's equation in vacuum Ric $= \Lambda g$ with cosmological constant $\Lambda = 3/R_{DS}^2$. Our universe is believed to be entering a de Sitter phase with R_{DS} around 12 billion light-years. We choose units in which $R_{DS} = 1$.[2]

The time-like geodesics in $\mathcal{DS}$ are the components of its intersections with hyperplanes through the origin of $\mathcal{M}^5$ of slope steeper than $45°$. The null geodesics of $\mathcal{DS}$ are the components of the intersections with hyperplanes through the origin of slope $45°$; note that they are null geodesics of $\mathcal{M}^5$.

Typical pairs of time-like geodesics in $\mathcal{DS}$ separate exponentially in both forwards and backwards time. Indeed the time-like geodesic flow is Anosov [70]. Exceptionally, pairs of time-like geodesics may converge together in backward time or in forward time.

We consider the null geodesics from a time-like emitter geodesic e to a time-like receiver geodesic r. By an isometry of $\mathcal{DS}$ we can bring the receiver geodesic to the form $x_0 = \sinh t, x_1 = \cosh t, x_j = 0$ for $j = 2, 3, 4$, with proper time t. The emitter geodesic can be expressed as $e = Mr$ for some future-preserving isometry M of $\mathcal{DS}$, equivalently, a linear isometry of $\mathcal{M}^5$. We parametrise the emitter geodesic by its proper time u, the image of t under M.

Since the null geodesics in $\mathcal{DS}$ are null in $\mathcal{M}^5$, the set of pairs (t, u) for which there is a future-pointing null geodesic from u on e to t on r is given by

(F.1) $- (a \sinh u + b \cosh u) \sinh t + (c \sinh u + d \cosh u) \cosh t = 1,$

with $a \sinh u + b \cosh u < \sinh t$, where $\begin{bmatrix} a & b \\ c & d \end{bmatrix}$ is the top 2×2 block of the matrix representing M. There are constraints on the values of a, b, c, d

---

[2]The notation used in this appendix differs slightly from that used in Section 9.3 and Appendix B where a is used for the de Sitter radius — a symbol used here for the top-left matrix entry, see below.

for them to come from an isometry matrix, namely

$$(ab - cd)^2 \leq (a^2 - c^2 - 1)(b^2 - d^2 + 1),$$

both factors on the right are non-negative, and $a \geq 1$.

Condition (F.1) can be written conveniently in terms of $T = e^t$ and $U = e^u$ as

$$-ATU + BT/U + CU/T - D/TU = 2,$$

with

(F.2) $2A = a + b - c - d$

(F.3) $2B = a - b - c + d$

(F.4) $2C = a + b + c + d$

(F.5) $2D = a - b + c - d,$

which are all non-negative. This has the causal solution

$$T = \frac{U + \sqrt{BD + (1 - BC - AD)U^2 + ACU^4}}{B - AU^2}$$

for $U < \sqrt{B/A}$. Equivalently we can write the emitter time as a function of receiver time. Each is monotone increasing in the other.

If $D, B, A \neq 0$ there is a first time t^* at which the emitter becomes visible, given by $T = \sqrt{D/B}$. Thus there is a sudden start to seeing the emitter, just as for gamma-ray bursts. We see its infinite past in a short interval of receiver time t. As $t \to +\infty$, we see the emitter up to a last emitter time u^* given by $U = \sqrt{B/A}$, but not beyond. See Figure F.1.

In the exceptional case $D = 0$ (e and r backward asymptotic) then $t^* = -\infty$; similarly if $A = 0$ (forward asymptotic) then $u^* = +\infty$. Finally, if $B = 0$ (e past asymptotic to the antipodal geodesic to r) then both $t^* = -\infty$ and $u^* = +\infty$.

Weyl [102] abhorred the idea that an object might suddenly become visible, so hypothesised that all emitter geodesics are backward asymptotic to ours. This eliminates, however, precisely the case we believe to be important for gamma-ray bursts.

We now study the redshift and intensity of the received light.

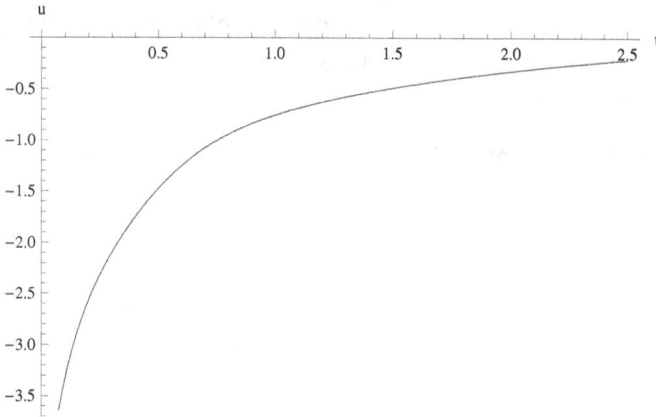

Figure F.1: Emitter time u as a function of receiver time t for a typical pair of time-like geodesics in $\mathcal{DS}$; the origins of receiver and emitter time have been shifted to t^*, u^* respectively.

The redshift z of an emitter relative to a receiver is defined by

$$1 + z = \frac{dt}{du} = \frac{U}{T} \frac{dT}{dU}.$$

An emitter frequency ω_e is transformed to a received frequency $\omega_r = \omega_e/(1+z)$. In the generic case $D, B, A > 0$, the redshift goes monotonically from -1 at t^* to $+\infty$ as $t \to +\infty$. Thus at its first appearance, the emitter is seen infinitely blue-shifted. Whatever it emits is seen as even higher frequency electromagnetic waves than gamma rays. If we assume the emitter spectrum is roughly constant in emitter time, then as receiver time advances, the received light descends through gamma rays to X-rays, visible and microwaves to radio waves, just as for gamma-ray bursts. For short time after the first appearance we have the asymptotic relation

(F.6) $1 + z \sim t - t^*.$

The emitter remains blue-shifted up to the time defined by $UT = \sqrt{D/A}$. The duration t_B of receiver time for which the emitter is seen blue-shifted comes out to

$$t_B = \frac{1}{2} \log \frac{1 + \sqrt{AD} + \sqrt{1 + 2\sqrt{AD} + AD - BC}}{\sqrt{AD}},$$

which provides a natural measure of the duration of the burst. In exceptional cases, z goes from 0 to $+\infty$ (backward asymptotic), or -1 to 0 (forward asymptotic) or jumps across 0 (intersecting geodesics).

The received flux Φ is related to the emitted power P per unit solid angle by [82]

$$\Phi = \frac{P}{((1+z)\rho)^2},$$

where ρ is called the "corrected luminosity distance", which accounts for the geometric expansion of the bundle of rays leaving a point on the emitter. In de Sitter space, ρ is given by the change in affine parameter along the null geodesic, scaled to correspond to elapsed time in the emitter frame initially. This yields

(F.7) $$\rho = 1 - (\frac{C}{T} - AT)U.$$

In the generic case $A, B, D > 0$, ρ starts from 1 (the de Sitter radius) at t^* and goes to $+\infty$ as $t \to +\infty$. Thus if the emitter power $P > 0$ at $u = -\infty$, the factor $(1+z)^2$ makes the received flux infinite initially. Even more, the received energy per unit area diverges for any receiver time interval including t^*, because of (F.6). Figure F.2 shows an example for constant P. In reality, we should expect P to be integrable as a function of emitter time u, thus the received flux is not infinite initially nor is the received energy infinite. Yet both may be extremely large, just as for gamma-ray bursts.

Note that ρ decreases initially if $BC > AD$ (put $T = \sqrt{D/B}$ in (F.7)). An example is shown in Figure F.3. This leads to an enhancement of the received flux. Indeed the received energy in time interval (t^*, t) can be written as

$$\int_{-\infty}^{u(t)} \frac{1}{\rho^3} \left(1 + \frac{D}{T}e^{-u} - \frac{C}{T}e^u \right) P(u) \, du.$$

The regime $BC \gg AD$ corresponds to that of short blue-shift period t_B. Thus we see that the brightest emitters are those with the shortest blue-shift period. This fits another feature of gamma-ray bursts, namely that those observed are very short compared to the de Sitter timescale.

To study the received flux further, it is convenient to apply isometries to reduce the generic case to $a = \cosh\phi$, $b = c = 0$, $d = \cos\theta$, for $\phi \geq 0$

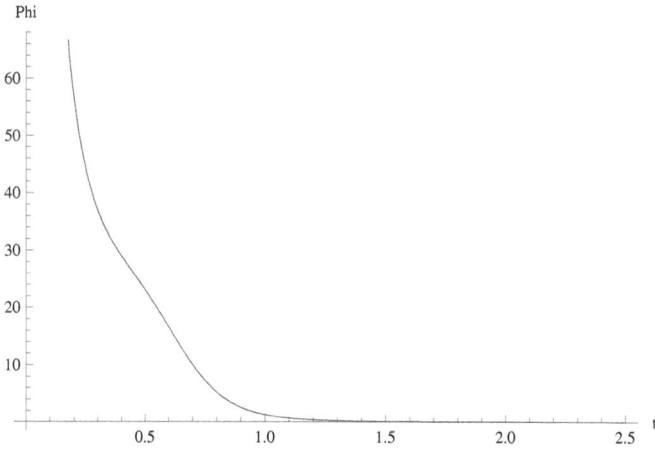

Figure F.2: An example of received flux Phi as a function of receiver time t since t^*, assuming constant emitter power P.

and $\theta \in [0, \pi]$. Then $A = D = (a - d)/2$ and $B = C = (a + d)/2$. The null geodesic condition reduces to

$$-a \sinh u \sinh t + d \cosh u \cosh t = 1,$$

and the redshift is given by

$$1 + z = \frac{d \tanh u - a \tanh t}{a \tanh u - d \tanh t}.$$

The blue-shift period for this reduced case can be written

(F.8) $$t_B = \log \frac{\sqrt{a + 1} + \sqrt{1 - d}}{\sqrt{a - d}},$$

and is plotted in Figure F.4.

We see that the shortest blue-shift periods are for ϕ large and θ near 0. Figure F.5 shows some light curves for short blue-shift periods. For the plots, we shifted the origin of t to $t^* = -\operatorname{arctanh}\frac{d}{a}$. Notice that a second hump occurs in some cases; this is due to $(1 + z)\rho$ coming to a local minimum. Such a second hump is a feature of many observed gamma-ray bursts (eg figures in [74]). For the reduced case, ρ can be written as

$$\rho = a \sinh t \cosh u - d \cosh t \sinh u,$$

and we calculate there is a second hump iff $d > \sqrt{8}/3$.

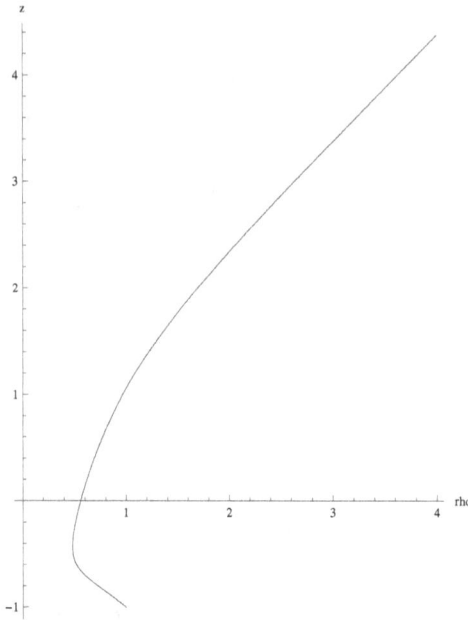

Figure F.3: HLSW diagram of redshift z against corrected luminosity distance rho for one emitter throughout its visible life.

Observed light curves are more complicated than ours, but one possible explanation is that the emitter power varies with emitter time, and any variations are compressed into a short interval of receiver time. We describe another possible contribution to the variability near the end of the paper.

We can also predict the distribution of durations. For definiteness, we use the blue-shift period as our measure of duration. We propose that the natural distribution for emitter geodesics in de Sitter space is invariant under isometries. This implies that the distribution on our two-parameter space of (ϕ, θ) is proportional to $\sinh^2 \phi \, \sin \theta \, d\phi \, d\theta$ (the distribution is non-normalisable). This can be written as $\sqrt{a^2 - 1} \, da \, dd$. Using (F.8) we obtain that the natural distribution for t_B is asymptotically $\frac{16}{3} t_B^{-5} \, dt_B$ for t_B small. So the natural distribution is heavily skewed to short blueshift period, which again fits well with observations of gamma-ray bursts. The difficulty is to explain why the observed density of durations (eg T_{90} in [62]) decreases for durations less than 20 seconds, but this could be because the

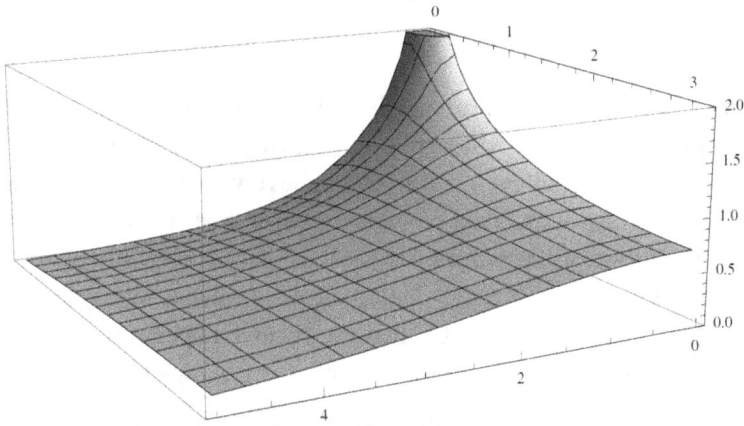

Figure F.4: Blue-shift period t_B as a function of ϕ (on the bottom axis) and θ (on the top axis) for the reduced family.

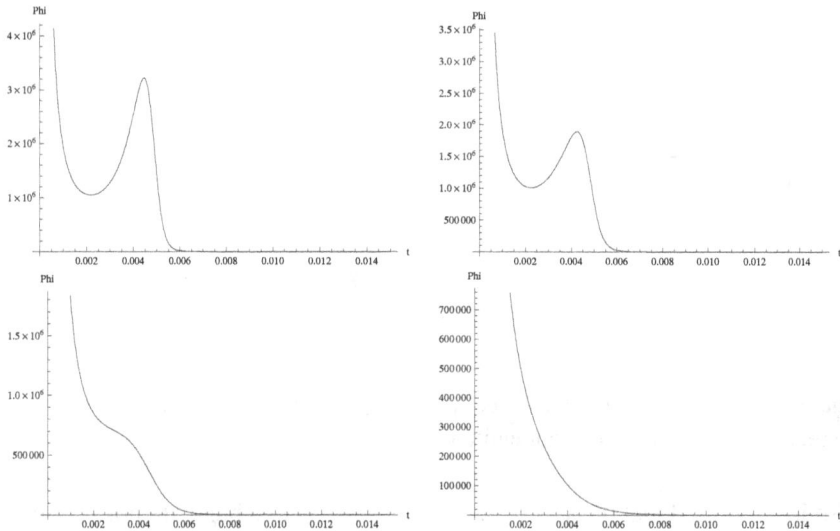

Figure F.5: Some light curves for constant emitters in $\mathcal{DS}$ with short blue-shift period

relevant part of space-time deviates a lot from $\mathcal{DS}$ or there are few emitters in the region corresponding to $t_B < 20$ seconds.

F.3   Critique

A common criticism of our proposal is that the observed bursts have non-thermal spectrum. There is no great reason to suppose that the emitter spectrum is thermal, but even if it is, we believe we can explain the non-thermal observations as an effect of averaging over time. Photon count rates are often very low, of the order of at most 10 per second, so to estimate a spectrum observations are averaged over a significant interval of time. If the emitter has temperature Θ then the received spectrum is thermal with temperature $\Theta/(1 + z)$. In our model, $1 + z$ varies rapidly initially. Averaging the received flux over a time interval produces spectra like that of Figure F.6, which agree well with observations like those of [48].

photonfluxperunitenergy

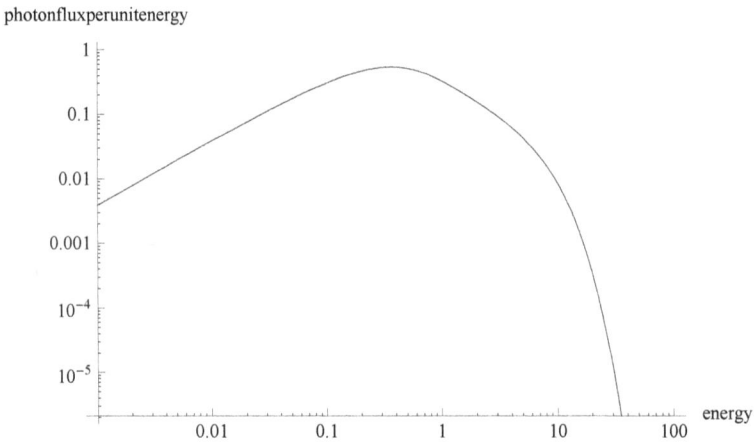

Figure F.6: Time-averaged spectrum of photons per unit time, area and energy, for the received flux from a constant emitter in $\mathcal{DS}$.

Also it is reported that the spectra at different stages of a burst are not simply Lorentz-boosted versions of each other. We have no problem with that, however, because it is perfectly natural that the emission spectrum (as well as the power) vary non-trivially during the emitter's life-time. We see the early history of the emitter compressed into a short interval of receiver time, so any such variations are accentuated.

Another criticism is that it is claimed that many gamma-ray bursts are associated with a distant galaxy which is in fact receding from us. This is done on the basis of searching for potential host galaxies immediately after a gamma-ray burst is observed, or detecting red-shifted absorption bands in the afterglow. We think this association may often be spurious. The gamma-ray burst could be from an emitter way beyond the purported host galaxy and which is approaching us rapidly. The absorption could indeed be by gas in an intermediate galaxy, but that does not imply the emitter is in that galaxy.

The most serious criticism is that our universe is believed to be nothing like $\mathcal{DS}$ in the past. It is said to contain sufficient matter and radiation to have made it collapse to a finite-time singularity in the past. We will not address the case for the Big Bang in this paper, but first we note that $\mathcal{DS}$ is not so far from the standard ΛCDM model. $\mathcal{DS}$ contains a flat Friedmann space-time with scale factor $S(t) = e^{t/R_{\mathcal{DS}}}$, which was in fact de Sitter's original space [92], namely the projection to space-time of the unstable manifold of a given time-like geodesic, so this part looks like an expanding universe, albeit going back to time $-\infty$ rather than a finite-time singularity. The key feature that seems to have been ignored since Weyl is that flat Friedmann space-times may be geodesically incomplete in other ways than a Big Bang. Weyl completed de Sitter's space in the way we presented it in this paper. In contrast to Weyl, however, we see no reason why objects should not suddenly become visible to us. Indeed, as we recall shortly, objects suddenly become visible in conventional Friedmann models. Nevertheless, we must examine how much of our mechanism survives deviations from $\mathcal{DS}$.

Small deviations of the metric from that of $\mathcal{DS}$ produce qualitatively the same time-like geodesic flow, because of the structural stability of Anosov systems. This means there is a near-identity homeomorphism taking time-like geodesics of any C^2-small perturbation of the metric to those of $\mathcal{DS}$. The proof does not extend to null geodesics, however, so there could be qualitative changes in the set of null geodesics connecting an emitter to a receiver. A suggested example is sketched in Figure F.7(a), corresponding to a swallowtail pleat in the forward light-cone of the emitter passing over the receiver. The cusps in the (u, t)-relation produce infinite intensity like $|t - t_0|^{-1/2}$ but the singularity is integrable, and the tangent to the cusp has slope in $(0, \infty)$ so there is no exceptional red or blue shifting. This is a relativistic version of the twinkling of stars. It is estimated that there

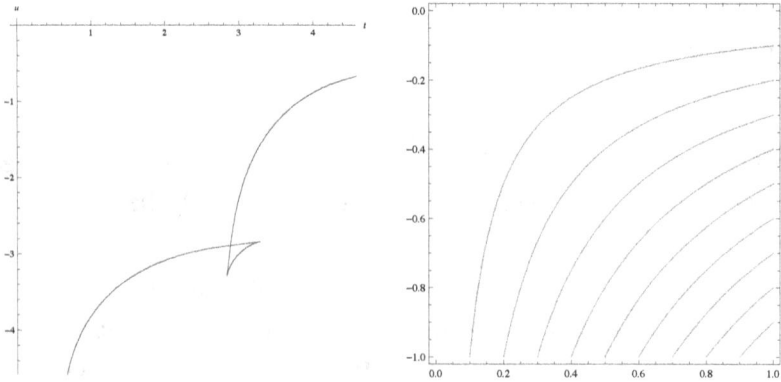

Figure F.7: Perturbed (u, t) relations with (a) two cusps, (b) an infinite sequence of branches.

are around 10^{22} caustics in our backward light-cone [41]. Each caustic of the emitter light-cone crossed by the receiver creates a cusp in the (u, t) diagram. So we should expect many cusps.

An alternative deformation of the (u, t) curve is to make a fold at an earlier time than t^*; then a precursor will be observed before the main burst, as in some observed cases.

If large perturbation from de Sitter space is considered then larger effects can be expected. For example, on introducing a Schwarzschild black hole, as in Kottler space, the (u, t)-diagram gains an infinite series of curves, corresponding to light paths making different numbers of turns around the black hole, as sketched in Figure F.7(b). The added travel time per turn in the black hole's frame is asymptotically $6\pi\sqrt{3}M$ for black hole mass M. Successive curves are presumably fainter.

Let us turn to Friedmann universes, those with metric $ds^2 = -dt^2 + S(t)^2 |dx|^2$ for some scale factor $S(t) > 0$, and suppose there is a Big Bang, ie $S(t)$ is defined for $t > 0$ only and $S(t) \to 0$ as $t \to 0$. If we are on a HLSW flow line $x = $ constant then we see every time-like geodesic redshifted initially. There is a first time $t^* > 0$ we begin to see it, corresponding to emitter time 0. We see it infinitely redshifted, unless its velocity is directly towards us when it is just finitely redshifted. The calculations are in [70]. If one allows cases with infinite past, however, like the case $S(t) = e^t$ for which the Friedmann universe is half of de Sitter space, and if emitter or receiver is not on the HLSW flow then our scenario for gamma-ray bursts occurs.

We believe there is room in between de Sitter space and big-bang universes for our mechanism for gamma-ray bursts to apply.

F.4 Final remark

We conclude by remarking that all emitters in de Sitter space except those converging to us in forwards time exhibit an asymptotic HLSW law $z \sim \rho$ for large positive time. Those on our unstable manifold do so exactly, but to obtain a good fit there is no need to require all visible matter to be converging together in backwards time.

Bibliography

[1] *CMB temperature map,*
http://map.gsfc.nasa.gov/media/121238/index.html

[2] *COBE satellite image,*
http://lambda.gsfc.nasa.gov/product/
cobe/cobe_images/cobeslide10.jpg

[3] *COBE satellite image: bulge,*
http://lambda.gsfc.nasa.gov/product/
obe/cobe_images/cobeslide11.jpg

[4] *Mathematica notebooks*, available from:
http://msp.warwick.ac.uk/~cpr/paradigm/Nb

[5] *Berkeley cosmology essays,*
http://cosmology.berkeley.edu/
Education/Essays/galrotcurve.htm

[6] *European Southern Observatory*, http://www.eso.org/public/

[7] *The Hubble site*, http://hubblesite.org

[8] *The Hubble Ultra-Deep Field*, download the 61Mb jpg from:
http://hubblesite.org/image/1457/news_release/2004-07

[9] *Nature book review, Ann Finkbeiner, The debated legacy of Einstein's first wife*, Nature 567, 28–29 (2019)
https://www.nature.com/articles/d41586-019-00741-6

[10] *Nature, News item, April 2009,*
http://www.nature.com/news/2009/
090401/full/news.2009.225.html

[11] *Nature, News item, October 2018*, Belgian priest recognized in Hubble-law name change,
https://www.nature.com/articles/d41586-018-07234-y

[12] *Physics.org, News item, April 2011*,
http://phys.org/news/
2011-04-galaxies-born-earlier-video.html

[13] *Quanta magazine: Kevin Hartnett: Abstractions blog, March 2018*
How Einstein lost his bearings, and with them, general relativity

[14] *ScienceNews item, Emily Conover, Sep 14 2019:* Debate over the universe's expansion rate may unravel physics. Is it a crisis?
https://www.sciencenews.org/article/
debate-universe-expansion-rate-hubble-constant-crisis

[15] *Stanford Encyclopedia of Philosophy*, Article on Hermann Weyl, section on cosmology,
https://plato.stanford.edu/entries/weyl/#RelCosWeyPos

[16] *The supernova cosmology project*, Lawrence Berkeley Laboratory, see:
http://panisse.lbl.gov

[17] *The high-Z supernova search*, see
http://www.cfa.harvard.edu/supernova//

[18] Wikimedia image: (6,4,2)-tesselation of the hyperbolic plane
https://commons.wikimedia.org/wiki/File:Hyperbolic_
domains_642.png

[19] Wikipedia article: *Aristarchus of Samos*,
https://en.wikipedia.org/wiki/Aristarchus_of_Samos

[20] Wikipedia article: *Nicolaus Copernicus*,
https://en.wikipedia.org/wiki/Nicolaus_Copernicus

[21] Wikipedia article: *Cosmic age problem*,
https://en.wikipedia.org/wiki/Cosmic_age_problem

[22] Wikipedia article: *Cosmic microwave background*,
https://en.wikipedia.org/wiki/Cosmic_microwave_
background

[23] Wikipedia article: *Globular clusters*,
http://en.wikipedia.org/wiki/Globular_cluster

[24] Wikipedia article: *Interstellar medium*,
https://en.wikipedia.org/wiki/Interstellar_medium

[25] Wikipedia article: *Michelson{Morley experiment*,
https://en.wikipedia.org/wiki/Michelson-Morley_
experiment

[26] Wikipedia article: *Philolaus*,
https://en.wikipedia.org/wiki/Philolaus

[27] Wikipedia article: *Relativity priority dispute*,
https://en.wikipedia.org/wiki/Relativity_priority_
dispute

[28] Wikipedia article: *General Relativity priority dispute*,
https://en.wikipedia.org/wiki/General_relativity_
priority_dispute

[29] Wikipedia article: *Sachs-Wolfe effect*,
https://en.wikipedia.org/wiki/Sachs-Wolfe_effect

[30] Wikipedia article: *Wien's displacement law*,
https://en.wikipedia.org/wiki/Wien's_displacement_law

[31] **H Arp**, *Sundry articles*, available at:
http://www.haltonarp.com/articles

[32] **G Berkeley**, *The principles of human understanding* (1710), *De Motu* (1726)

[33] **K C Begeman**, *H1 rotation curves of spiral galaxies*, Astron Astrophys 223 (1989) 47–60

[34] **A E Broderick, A Loeb, R Narayan**, *The event horizon of Sagittarius A**, Astrophys J. 701 (2009) 1357–1366

[35] **J Binney, M Merrifield**, *Galactic Astronomy*, Princeton UP (1998)

[36] **J Binney, S Tremaine**, *Galactic Dynamics*, Second Edition, Princeton UP (2008)

[37] **H Bondi, T Gold, F Hoyle**, *Black giant stars*, Observatory 75 (1955) 80

[38] **C A Collins, et al.**, *Early assembly of the most massive galaxies*, Nature 458 (2009) 603–606

[39] **A S Eddington**, *On the instability of Einstein's spherical worlds*, Mon Not Roy Astron Soc 90 (1930) 668–678

[40] A Einstein, *Zur Electrodynamik bewegter Körper*, Annalen der Physik 17 (1905) 891–921

[41] G F R Ellis, B A C C Bassett, P K S Dunsby, *Lensing and caustic effects on cosmological distances*, Class Quantum Grav, 15 (1998) 2345–61

[42] C W F Everitt et al, *Gravity Probe B: Final Results of a Space Experiment to Test General Relativity*, Phys. Rev. Lett. 106, 221101

[43] R A Flammang, K S Thorne, A N Zytkow, *Stationary spherical accretion into black holes*, Mon Not Roy Astron Soc 194 (1981) 475–484

[44] A Friedmann, *Über die Krümmung des Raumes*, Zeitschrift für Physik A, 10 (1922) 377–386

[45] P Galianni, E M Burbidge, H Arp, V Junkkarinen, G Burbidge, Stefano Zibetti, *The discovery of a high redshift X-ray emitting QSO very close to the nucleus of NGC 7319*, arXiv:astro-ph/0409215

[46] G W Gibbons, S W Hawking, *Cosmological event horizons, thermodynamics and particle creation*, Phys Rev D 15 (1977) 2738–2751

[47] J L Greenstein, M Schmidt, *The quasi-stellar sources radio sources 3C48 and 3C273*, Astrophys J. 140 (1964) 1–34

[48] A Goldstein et al, *The Fermi GBM gamma-ray burst spectral catalog: the first two years*, Astrop J Suppl, 199 (2012) 19–45

[49] J T Giblin, D Marolf, R Garvey, *Spacetime embedding diagrams for spherially symmetric black holes*, General Relativity and Gravitation 36, (2004) 83–99

[50] S Gillessen et al, *Monitoring stellar orbits around the Massive Black Hole in the Galactic Center*, Astrophys J 692 (2009) 1075 DOI:10.1088/0004-637X/692/2/1075

[51] M G Hauser, E Dwek, *The cosmic infrared background: measurements and implications*, Annu Rev Astron Astrophys 39 (2001) 249–307

[52] M R S Hawkins, *On time dilation in quasar light curves*, Mon Not Roy Astron Soc, 405 (2010) 1940–6

[53] F Hoyle, G Burbidge, *Anomalous redshifts in the spectra of extragalactic objects*, Astron. Astrophys. 309 (1996) 335–344

[54] F Hoyle, G Burbidge, J V Narlikar, *A quasi-steady state cosmological model with creation of matter*, Astrophys J 410 (1993) 437–457 DOI:10.1086/172761

[55] **F Hoyle, G Burbidge, J V Narlikar**, *A different approach to cosmology*, CUP, Cambridge (2000)

[56] **F Hoyle, A Fowler**, Nature 213 (1964) 217

[57] **F Hoyle, N C Wickramasinghe**, *Proofs that life is cosmic*, Mem. Inst. Fund. Studies Sri Lanka (1982)

[58] **E Hubble**, *A relation between distance and radial velocity among extragalactic nebulae*, PNAS 15 (1929) 168–173

[59] **E Hubble, M L Humason**, *The Velocity-Distance Relation among Extra-Galactic Nebulae*, Astrophys J 74 (1931) 43–80

[60] **M Janssen**, *The Einstein-De Sitter-Weyl-Klein Debate*, from: Schulmann et al, "Vol. 8 of The Collected Papers of Albert Einstein, covering correspondence during period 1914–1918", (1998) 351–357

[61] **D Kastor, J Traschen** (1993) *Cosmological multi-black-hole solutions*, Phys Rev D 47 (1993) 5370–5

[62] **C Kouveliotou, C A Meegan, G J Fishman, N P Bhat, M S Briggs, T M Koshut, W S Paciesas, G N Pendleton**, *Identification of two classes of gamma-ray bursts*, Astrop J, 413 (1993) L101–4

[63] **C Leipski, K Meisenheimer**, *The dust emission of high-redshift quasars*, J. Phys.: Conf. Ser. 372 (2012) 012037

[64] **G Lemaître**, *Un Univers homogène de masse constante et de rayon croissant rendant compte de la vitesse radiale des nébuleuses extra-galactiques*, Ann Soc Sci Bruxelles A47 (1927) 49–59

[65] **G Lemaître**, *L'univers en expansion*, Ann Soc Sci Bruxelles A53 (1933) 51–85,

[66] **M Longair**, Galaxy Formation, Springer Astron–Astrophys Library, 2nd Ed (2008)

[67] **M Lopez-Corredoira, C M Gutierrez**, *Two emission line objects with $z > 0.2$ in the optical filament apparently connecting the Seyfert galaxy NGC 7603 to its companion*, arXiv:astro-ph/0203466v2

[68] **M Lopez-Corredoira, C M Gutierrez**, *Research on candidates for noncosmological redshifts*, arXiv:astro-ph/0509630v2

[69] **E Mach**, *The science of mechanics: a critical and historical account of its development*, Translation from the original German by T J McCormack, Open Court Publishing Co (1893) (page numbers in the text refer to the sixth American edition (1960) Lib. Congress cat. no. 60-10179)

[70] **R S MacKay, C Rourke**, *Natural flat observer fields in spherically-symmetric space-times*, J. Phys. A: Math. Theor. 48 (2015) 225204, available at:
http://msp.warwick.ac.uk/~cpr/paradigm/escape-Jan2015.pdf

[71] **R MacKay, C Rourke**, *Natural observer fields and redshift*, J Cosmology 15 (2011) 6079–6099
http://msp.warwick.ac.uk/~cpr/paradigm/
redshift-nat-final.pdf

[72] **R S MacKay, C Rourke**, *Are gamma-ray bursts optical illusions?* Palestinian J Math 5(Spec.1) (2016) 175–197, available at:
http://msp.warwick.ac.uk/~cpr/paradigm/GammaRayBursts.pdf

[73] **D M Meier**, *Black hole astrophysics: the engine paradigm*, Springer (2009)

[74] **P Meszaros** *Gamma-ray bursts*, Rep Prog Phys, 69 (2006) 2259–2321

[75] **F C Michel**, *Accretion of matter by condensed objects*, Astrophys and space sci 15 (1972) 152–160

[76] **C W Misner, K S Thorne, J A Wheeler**, *Gravitation*, Freeman (1973)

[77] **U Moschella**, The de Sitter and anti-de Sitter sightseeing tour, Séminaire Poincaré 1 (2005) 1–12, available at:
http://www.bourbaphy.fr/moschella.pdf

[78] **R Narayan, J E McClintock** *Advection-Dominated Accretion and the Black Hole Event Horizon*, arXiv:0803.0322

[79] **J H Oort, F J Kerr, G Westerhout**, *The galactic system as a spiral nebula*, MNRAS 118 (1958) 379

[80] **M D'Onofrio, J W Sulentic, P Marziani**, (Editors), *Fifty years of quasars*, Astrophys & Space Sci Lib 386, Springer (2012)

[81] **C O'Raifeartaigh, S Mitton**, *A new perspective on steady-state cosmology: from Einstein to Hoyle*, arXiv:1506.01651

[82] **Perlick V**, *Gravitational Lensing from a Spacetime Perspective*, Living Reviews Relativity, 7 (2004) 9

[83] **M J Reid et al**, *The Proper Motion of Sagittarius* A*. *I. First VLBA Results*, AstroPhys J 524 (1999) 816 DOI:10.1086/307855

[84] **W Rindler**, *The Lense–Thirring effect exposed as anti-Machian*, Phys Lett A 187 (1994) 236–238

[85] **H P Robertson**, *Kinematics and world structure I,II,III*, Astroph. J 82 (1935) 284–301, 83 (1936) 187–201 and 257–271

[86] **C Rourke**, *Intrinsic redshift in quasars*, available at: http://msp.warwick.ac.uk/~cpr/paradigm/ hawkins-time-dilation.pdf

[87] **C Rourke**, *Uniqueness of spherically-symmetric vacuum solutions to Einstein's equations*, notes available at: http://msp.warwick.ac.uk/~cpr/paradigm/uniqueness.pdf

[88] **C Rourke, R Toala Enriques, R S MacKay**, *Black holes, redshift and quasars*, draft preprint, available at: http://msp.warwick.ac.uk/~cpr/paradigm/quasars.pdf

[89] **D Sciama**, *On the origin of inertia*, Mon. Not. Roy. Astrom. Soc. 113 (1953) 34–42

[90] **D Sciama, P C Waylen, R C Gilman**, *Generally covariant integral formulation of Einstein's field equations*, Phys. Rev. 187 (1969) 1762–1766

[91] **R E Schild, D J Leiter, S L Robertson** *Observations Supporting the Existence of an Intrinsic Magnetic Moment inside the Central Compact Object within the Quasar Q0957+561*, Astron J, 132 (2006) 420 DOI:10.1086/504898

[92] **W de Sitter**, *On the relativity of inertia*, Kon Ned Akad Wet Proc, 19 II (1917) 1217–25.

[93] **W de Sitter**, *On the curvature of space*, Kon Ned Akad Wet Proc, 20, (1918) 229–243

[94] **V M Slipher**, *Radial velocity observations of spiral nebulae*, The Observatory 40 (1917) 304–306

[95] **A Stockton**, *The nature of QSO redshifts*, Astrophys J. 223 (1978) 747–757

[96] **Y Sofue, V Rubin**, *Rotation curves of galaxies*, Annu Rev Astron Astrophys 39 (2001) 137–174 arXiv:astro-ph/0010594v2

[97] **H Thirring, J Lense**, *Über den Einfluss der Eigenrotation der Zen-
      tralkörper auf die Bewegung der Planeten und Monde nach der Einstien-
      schen Gravitationstheorie*, Phys Z. 19 (1918) 156–163

[98] **S Tang, S N Zhang**, *Evidence against non-cosmological redshifts of QSOs
      in SDSS data*, arXiv:0807.2641v2

[99] **K P Tod**, *Mach's Principle revisited*, Gen. Rel. Grav. 26 (1994) 103–111

[100] **D E Vandem Berk et al**, *Composite quasar spectra from the Sloan digital
      sky survey*, Astronomical

[101] **A G Walker**, *On Milne's theory of world-structure*, Proc London Math
      Soc (2)42 (1937) 90–127

[102] **H Weyl**, *Zur allgemeinen Relativitätstheorie*, Phys Zeit, 24 (1923) 230–2

[103] **H Weyl**, *Raum, Zeit, Materie*, 5 edn, revised, Springer, Berlin (1923)

[104] **K Z Win**, *Ricci tensor of diagonal metric*, arXiv:gr-qc/9602015

[105] **E L Wright**, *Lyman Alpha Forest*, web site at
      http://www.astro.ucla.edu/~wright/
      Lyman-alpha-forest.html

[106] **E C Zeeman**, *Causality Implies the Lorentz Group*, J Math Phys 5, 490
      (1964)

[107] **F Zwicky**, *Die Rotverschiebung von extragalaktischen Nebeln*, Helvetica
      Physica Acta 6 (1933) 110–127. See also Zwicky, F. (1937). *On the Masses
      of Nebulae and of Clusters of Nebulae*, Astrophysical J 86 (1937) 217
      DOI:10.1086/143864

Index

SERIES ON KNOTS AND EVERYTHING
ISSN: 0219-9769

Editor-in-charge: Louis H. Kauffman *(Univ. of Illinois, Chicago)*

The Series on Knots and Everything: is a book series polarized around the theory of knots. Volume 1 in the series is Louis H Kauffman's Knots and Physics.

One purpose of this series is to continue the exploration of many of the themes indicated in Volume 1. These themes reach out beyond knot theory into physics, mathematics, logic, linguistics, philosophy, biology and practical experience. All of these outreaches have relations with knot theory when knot theory is regarded as a pivot or meeting place for apparently separate ideas. Knots act as such a pivotal place. We do not fully understand why this is so. The series represents stages in the exploration of this nexus.

Details of the titles in this series to date give a picture of the enterprise.

More information on this series can also be found at http://www.worldscientific.com/series/skae

www.ingramcontent.com/pod-product-compliance
Lightning Source LLC
Chambersburg PA
CBHW050550190326
41458CB00007B/1984